Chemical and Biological Processes in Fluid Flows

A Dynamical Systems Approach

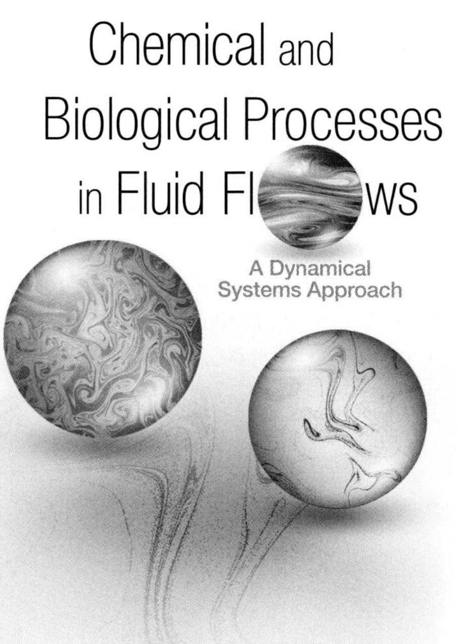

Chemical and Biological Processes in Fluid Flows

A Dynamical Systems Approach

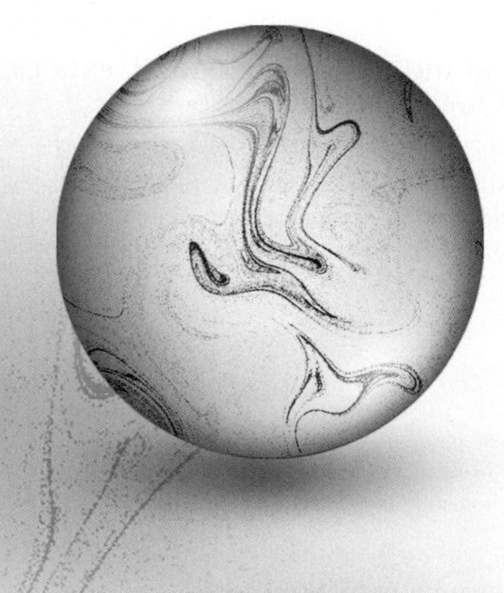

Zoltán Neufeld
University College Dublin, Ireland

Emilio Hernández-García
IFISC, Spanish Higher Research Council (CSIC) &
University of the Balearic Islands, Spain

Imperial College Press

Published by

Imperial College Press
57 Shelton Street
Covent Garden
London WC2H 9HE

Distributed by

World Scientific Publishing Co. Pte. Ltd.
5 Toh Tuck Link, Singapore 596224
USA office: 27 Warren Street, Suite 401-402, Hackensack, NJ 07601
UK office: 57 Shelton Street, Covent Garden, London WC2H 9HE

British Library Cataloguing-in-Publication Data
A catalogue record for this book is available from the British Library.

ISBN-13 978-1-86094-699-8
ISBN-10 1-86094-699-2

Printed in Singapore by B & Jo Enterprise Pte Ltd

To our wives, Gabriella and Roser,
for their unfaltering patience and unconditional support.

Z.N. and E.H.-G.

Contents

Preface

Chemically and biologically reacting flows

In any elementary chemistry lab, students intuitively learn that usual reactions proceed faster when the reactants, normally in aqueous solution, are vigorously stirred. The reason is the same than the one by which one stirs coffee and milk to get an homogenous drink: chemistry is about molecules colliding to recombine their atoms into new configurations. And no collision will be possible if the different molecules sit in different parts of a container. It seems obvious that *mixing* together the different components of the reacting fluid will improve the chances of molecular encounters, thus increasing reactivity. And mechanical stirring, i.e. producing irregular motion of the fluid containing the reactants, is an efficient way to mix them.

These apparently simple arguments have been found to be extremely difficult to formalize. The molecular contact needed for chemical reactions occurs at scales many orders of magnitude smaller than the ones at which mechanical energy is injected as stirring. The transfer of energy and inhomogeneities across scales is thus a major ingredient in the mixing process, and it is essentially tied to the very difficult problem of understanding and modelling turbulence. In addition to the complexity of the fluid dynamics, an extra layer of difficulty arises from the nonlinearity inherent to most chemical reactions. As a consequence of all of this, unexpected phenomena can occur. For example, one can find reactions (Epstein, 1995) in which changing the rate at which the solution is stirred can cause a transition from a state in which the concentrations remain stationary to one of periodic or even chaotic concentration oscillations; or to crystallization processes that yield sometimes all left-handed and sometimes all right-handed crystals; or to reactions that occur suddenly at random intervals, with mean reaction times depending on stirring. Reaction times can even be longer, not shorter, for increased stirring (Nagypál and Epstein, 1986).

Biological population dynamics can be thought also as a set of complex "chemical reactions" in which organisms, instead of molecules, experience interactions and transformations: biological entities replicate if they find adequate nutrients, predators annihilate preys, and resources are consumed. In addition to occurring in interesting laboratory settings, the interaction of biological dynamics and turbulent flows is an essential ingredient in the life cycle of planktonic organisms (Mann and Lazier, 1991), which are defined as the aquatic living beings unable to completely overcome the ambient ocean currents. Although planktonic species are much larger than molecules, they are also subjected to influences from a huge range of spatial scales, from the microbial ones to the planetary sizes characteristic of the large-scale ocean circulation. As in the case of chemistry, unexpected responses to fluid flow occur which are still far from being fully understood (Martin, 2003; Catalan, 1999; Peters and Marrasé, 2000). Other situations involving reactive fluids in geophysical-scale flows are typical in atmospheric chemistry, or convection in the Earth mantle.

The great complexity of the situations encountered when dealing with biological or chemical interactions under fluid mixing and dispersion has lead some authors to very pessimistic statements: "*... One should expect no general results that describe in all (or most) cases how 'the biology' modifies a spatial pattern that has arisen solely from the dispersal of species ...*". (Powell and Okubo, 1994). In the chemical context, Ottino (1994) states: "*From a theoretical viewpoint, mixing problems appear complex and unwieldy; from an applied viewpoint, it is easy to get lost in the complexities of particular cases without ever seeing the structure of the entire subject. Thus, generally speaking, mixing problems in nature and technology are attacked on a case-by-case basis. Specific experimental results, however, can rarely be extrapolated and there is a pressing need to generate more general results.*"

The aim of this book is to contribute to this last request. Without underestimating the diversity and complexity of the situations encountered, we believe that there are some common themes in many of them, which could serve as a guide to find general rules and broad enough methodologies. One of such unifying conceptual frameworks is the paradigm of *chaotic advection* or *Lagrangian chaos* (Aref, 1984, 2002), which allows to transfer to fluid dynamics problems the vast amount of knowledge which has been developed in the study of nonlinear dynamical systems. The power of this approach has been greatly enhanced by recent experimental and theoretical developments in Lagrangian approaches to fluid dynamics in general, and to turbulence in particular (Shraiman and Siggia, 2000; Falkovich et al., 2001), and by the observation of similarities between developed turbulence and low-dimensional chaotic dynamics when considered from the Lagrangian point of view (Bohr et al., 1998, Chap. 8).

The application of dynamical systems concepts to understanding reactions in flows has been developing during the last decade or so, and summaries of the most relevant results have already been compiled for example in the Chaos Focus Issue published by Toroczkai and Tél (2002), and notably, in the very good review by Tél et al. (2005). It is not our intention to repeat here the contents of these works. We noticed instead the need of integrating this recent work into the context of more traditional fluid dynamics and turbulence approaches. Also, the generality of the results is better appreciated after surveying general concepts in chemical, biological, and fluid dynamics. In consequence we have written a significant amount of introductory material on these subjects. We expect that the contents of these first chapters will be useful not only for establishing the adequate framework for the developments in the following ones, but also to provide the interested reader with a compilation of results which are now scattered in different monographs on nonlinear dynamics, fluid mechanics, nonlinear chemistry, and mathematical biology.

Plan of the book

After this introduction, we start (Chapter 1) with a summary of basic facts in fluid dynamics, such as the fundamental equations, the phenomenology and description of turbulence, and the peculiarities of two-dimensional flows. Chapter 2 is devoted to the fundamentals of *dispersion* and *mixing*, i.e. to the mechanisms that move around substances contained in the fluid, eventually homogenizing the different parts. Transport by advection, with a description of the Lagrangian point of view, and molecular, turbulent and anomalous diffusion processes are introduced. Later, advection and diffusion are considered in a sequence of flow-types of increasing complexity: two-dimensional steady, two-dimensional weakly time-dependent, three-dimensional steady, and turbulent. The phenomenon of chaotic advection (Sect. 2.5), one of the unifying themes in the following, is presented, and tools are given for its description. In particular we introduce Lyapunov exponents and a filament, or lamellar, model (Sect. 2.7.1) which turns out to be a powerful tool to analyze mixing and complex reactions in flows. The motion of suspended particles which do not follow exactly the fluid velocity (i.e. inertial particles) is also briefly described.

Chapter 3 is an overview of chemical and biological nonlinear dynamics. The kinetics of several types of reactions –first order, binary, catalytic, oscillatory, etc.– and of ecological interactions –predation, competition, birth and death, etc.– is described, nearly always within the framework of differential equations. The aim of this Chapter is to show that, despite the great variety of mechanisms and processes occurring, a few mathematical structures appear recurrently, and archetypical simplified models can be analyzed to understand whole classes of chemical or biological phenomena. The presence of very different timescales and the associated methodology of *adiabatic elimination* is instrumental in recognizing that.

The rich consequences of adding a diffusive mechanism of transport to chemical or biological activity are described in Chapter 4. Fisher waves and other types of fronts, excitable waves, Turing patterns, and other spatiotemporal phenomena produce striking structures which are observed in chemical and biological media. Understanding them is needed before addressing the additional impact that advection has on these systems.

After these introductory Chapters, we begin in Chapter 5 the study of situations in which reaction, advection, and diffusion act together. We address simplified interactions and flows, which are representative of the different situations previously enumerated. The first case considered is that of a binary reaction occurring sufficiently fast, so that the process becomes limited by the slow diffusive processes which mixes the different chemicals which react. The filament model of Sect. 2.7.1 allows a rather complete description of this situation both in simple and in complex flows. Chapter 6 studies chemical or biological processes which decay, at least locally, to a steady state. The linear regime of such decay can be also understood both in laminar as in turbulent flows, leading to filamental spatial structures with strong intermittency properties. This problem has also implications for the decay of two-dimensional turbulence (Sect. 6.5).

Chapter 7 considers reactions having some kind of autocatalytic behavior, so that one of the components tends to grow fuelled by itself. This is the standard multiplicative reproduction undergone by many biological organisms. Excitable, bistable, and other types of reactions usually have an autocatalytic step, and the impact of flow on these is also discussed here. Oscillations under homogeneous, inhomogeneous and stochastically perturbed situations are considered in Chapter 8.

The spectrum of interesting problems involving active substances in flows is too broad to be fully covered here, and many topics are not touched in this book. The final Chapter 9 summarizes some of the most important ones, giving suggestions for further reading.

Chapter 1

Fluid Flows

Flowing fluids are ubiquitous in Nature, from large scale atmospheric winds and oceanic currents to the circulation of blood or flow around microorganisms swimming in a liquid media. Fluid motion is also important in industrial processing and affects the motion of vehicles (cars, aircrafts etc.). In this chapter we briefly review the basic concepts and fundamental laws describing the motion of fluids. More details can be found in fluid dynamics textbooks (see, e.g. Batchelor (1967), Lamb (1932), Landau and Lifschitz (1959), Tritton (1988)).

In most cases one is interested in fluid flows at scales that are much larger than the distance between the molecules. The value of the molecular mean free path in air at room temperature and 1 atm of pressure is $\lambda = 6.7 \times 10^{-8}$ m and in water $\lambda = 2.5 \times 10^{-10}$ m. When the Knudsen number – defined as the ratio of the molecular mean-free-path to a characteristic length scale of the flow (e.g. the size of the smallest eddies) – is small, the fluid can be described as a continuous medium in motion. In this continuum approximation the flow can be characterized by the velocity field $\mathbf{v}(\mathbf{x}, t)$ representing the instantaneous velocity of infinitesimal fluid elements at time t and at position $\mathbf{x}$. Fluid elements represent small volumes of fluid that are much smaller than the smallest characteristic scale of the flow, but sufficiently large to contain a large number of molecules so that a well defined local velocity exists and molecular fluctuations can be neglected.

1.1   Conservation laws

The equations of fluid motion can be derived from the physical principles of conservation of mass, momentum and energy. The standard form of such conservation equations is

$$\frac{\partial A}{\partial t} + \nabla \cdot \mathbf{J} = S(\mathbf{x}, t) \ . \tag{1.1}$$

$A(\mathbf{x}, t)$ is the density of the quantity which is conserved, $\mathbf{J}(\mathbf{x}, t)$ is the flux, i.e. the amount of that quantity crossing the unit surface at location $\mathbf{x}$ per unit of time. $S(\mathbf{x}, t)$ is the production (or consumption) rate of the quantity A per unit of volume at location $\mathbf{x}$. When $S = 0$, the balance between the terms in the left-hand-side indicates that the quantity A changes locally just because it is moving from place to place, so that it is globally conserved.

The conservation of the mass of a fluid in a velocity field $\mathbf{v}(\mathbf{x}, t)$ is expressed by Eq. (1.1) for the fluid density (mass per unit of volume) field $\rho(\mathbf{x}, t)$, with $S = 0$. The only process moving the fluid mass is transport by the velocity $\mathbf{v}$, giving the advective flux $\mathbf{J} = \rho \mathbf{v}$. Thus, Eq. (1.1) becomes the continuity equation

$$\frac{\partial \rho}{\partial t} + \nabla \cdot (\rho \mathbf{v}) = 0 \ . \tag{1.2}$$

The second term can be written as a sum of two components: $\mathbf{v} \cdot \nabla \rho$ that represents changes in the local density due to the flow bringing-in fluid of different density from elsewhere, and $\rho \nabla \cdot \mathbf{v}$ representing compression or expansion of the fluid volume when the velocity field is convergent ($\nabla \cdot \mathbf{v} < 0$) or divergent ($\nabla \cdot \mathbf{v} > 0$).

Under normal conditions most fluids are not compressed much in a flow. In general, if the typical flow velocity (U) is much smaller than the speed of sound (c) in the medium ($c_{air} \approx 340$ m/s, $c_{water} \approx 1500$ m/s), i.e. the Mach number, Ma $\equiv U/c$, is small, then the fluid is essentially incompressible. In this case the velocity field is a divergence-free (solenoidal) vector field

$$\nabla \cdot \mathbf{v} = 0 \ , \tag{1.3}$$

implying that the fluid density at a fixed location can only change due to transport from other fluid regions of different density.

The rate of change of a physical field $f(\mathbf{x}, t)$ along the trajectory of a fluid element, $\mathbf{r}(t)$, is obtained by differentiating $f[\mathbf{x} = \mathbf{r}(t), t]$ as

$$\frac{\mathrm{D}f}{\mathrm{D}t} = \frac{\partial f}{\partial t} + \sum_i \frac{\partial f}{\partial x_i} \frac{\mathrm{d}x_i}{\mathrm{d}t} = \frac{\partial f}{\partial t} + (\mathbf{v} \cdot \nabla)f , \tag{1.4}$$

where the two components represent the rate of change in time at a fixed position and the change due to the motion of the fluid, respectively. The operator

$$\frac{\mathrm{D}}{\mathrm{D}t} = \frac{\partial}{\partial t} + \mathbf{v} \cdot \nabla \tag{1.5}$$

is called the Lagrangian or material derivative. Thus, for an incompressible flow the continuity equation states that the Lagrangian derivative of the density is zero, $\mathrm{D}\rho/\mathrm{D}t = 0$, meaning that the density is conserved along the path of the fluid elements. If initially the density is uniform in space then an incompressible flow maintains this uniform density.

The basic equation for the fluid velocity results from the conservation of momentum and can be obtained by applying Newton's law to a fluid element

$$\rho \frac{\mathrm{D}\mathbf{v}}{\mathrm{D}t} = \mathcal{F} + \nabla \cdot \boldsymbol{\sigma} . \tag{1.6}$$

Acceleration of the fluid elements is due to two types of forces, body forces (per unit of volume), $\mathcal{F}$, acting within the whole fluid element volume (e.g. gravitational or electromagnetic forces) and forces acting on the surface of the small fluid element representing interaction with the rest of the fluid. The surface forces per unit of volume can be written as the divergence of a stress tensor $\boldsymbol{\sigma}$.

The stress tensor can be expressed as the sum of an isotropic diagonal part, a normal stress described by a scalar pressure p, and a deviatory component or shear stress that appears as a result of internal friction between fluid layers moving relative to each other. In the simplest case of an incompressible Newtonian fluid the stress tensor is of the form

$$\boldsymbol{\sigma} = -p\mathbf{I} + \mu(\nabla \mathbf{v} + \nabla \mathbf{v}^T) , \tag{1.7}$$

where the shear stress is assumed to be proportional to the strain tensor that describes the infinitesimal deformation rates of the fluid.

The coefficient μ is the dynamic viscosity, a fluid property that characterizes the resistance of the fluid to shearing forces and $(\nabla \mathbf{v})_{ij} = \partial v_i / \partial x_j$ is the velocity gradient tensor. Using this form of the stress tensor one obtains the Navier-Stokes equation

$$\frac{\partial \mathbf{v}}{\partial t} + \mathbf{v} \cdot \nabla \mathbf{v} = -\frac{1}{\rho} \nabla p + \nu \nabla^2 \mathbf{v} + \mathcal{F} , \qquad (1.8)$$

where $\nu = \mu / \rho$ is the kinematic viscosity. Typical values are $\nu_{air} = 1.5 \times 10^{-5}$ m^2/s and $\nu_{water} = 10^{-6}$ m^2/s. Thus the velocity at a point fixed in space changes due to inertia, i.e. advection of the velocity field by itself, pressure differences, internal friction and body forces. The equation needs to be supplemented with boundary conditions expressing that there is no flow through the boundary of the domain, and the "no-slip" condition, i.e. the fluid velocity at a solid boundary should be equal to that of the boundary. There are some special cases (e.g. in large-scale ocean models) where the no-slip condition is relaxed to allow for unrestricted fluid motion along the boundary. For ideal inviscid fluids ($\nu = 0$) the term representing viscous dissipation is dropped and the corresponding equation is called the Euler equation.

The momentum equation (1.8) together with the additional constraint of incompressibility (1.3) fully define the motion of an incompressible fluid. They form a system of four scalar equations for four unknown functions: the three components of the velocity field and the pressure field. Note that there is no evolution equation for the pressure field, which is given implicitly through the incompressibility condition. This somewhat complicates the analysis and numerical solution of the Navier-Stokes equation.

One way to eliminate the pressure from the Navier-Stokes equation is by transforming it into an equation for the vorticity field defined as the curl of the velocity field, $\boldsymbol{\omega} = \nabla \times \mathbf{v}$, that characterizes the distribution and direction of the rotational motion in the flow. By taking the curl of both sides of (1.8) and using the vector identity $(\mathbf{a} \cdot \nabla)\mathbf{a} = (1/2)\nabla|\mathbf{a}|^2 + (\nabla \times \mathbf{a}) \times \mathbf{a}$ and that for any scalar field ψ, $\nabla \times \nabla \psi = 0$ one obtains

$$\frac{\partial \boldsymbol{\omega}}{\partial t} = \nabla \times (\mathbf{v} \times \boldsymbol{\omega}) + \nu \nabla^2 \boldsymbol{\omega} + \nabla \times \mathcal{F}. \qquad (1.9)$$

The nonlinear term can be further expressed as

$$\nabla \times (\mathbf{v} \times \boldsymbol{\omega}) = (\boldsymbol{\omega} \cdot \nabla)\mathbf{v} - (\mathbf{v} \cdot \nabla)\boldsymbol{\omega} + \mathbf{v}(\nabla \cdot \boldsymbol{\omega}) - \boldsymbol{\omega}(\nabla \cdot \mathbf{v}) \,, \quad (1.10)$$

where the last two terms vanish because of $\nabla \cdot \boldsymbol{\omega} = \nabla \cdot (\nabla \times \mathbf{v}) = 0$ and incompressibility, $\nabla \cdot \mathbf{v} = 0$. Thus the dynamics of the vorticity field is described by the equation

$$\frac{\partial \boldsymbol{\omega}}{\partial t} + \mathbf{v} \cdot \nabla \boldsymbol{\omega} = \boldsymbol{\omega} \cdot \nabla \mathbf{v} + \nu \nabla^2 \boldsymbol{\omega} + \nabla \times \mathcal{F}. \quad (1.11)$$

This shows how the vorticity of a fluid is affected by the forcing and by the diffusion of vorticity. In addition, the first term on the right-hand-side representing "vortex stretching" also affects the vorticity field and the intensification of vorticity by the stretching of vortex tubes allows the transfer of vorticity and energy from large to smaller scales.

From the energetic point of view the work of the external forces produces fluid motion with a certain kinetic energy that is transformed into heat by internal friction. The energy balance can be obtained by multiplying the Navier-Stokes equation (1.8) by $\mathbf{v}$ and integrating over the whole domain (an operation denoted by brackets $\langle ... \rangle$). The contribution of the advective and pressure terms vanish for periodic or no-slip boundary conditions and we obtain

$$\frac{\mathrm{d}E}{\mathrm{d}t} \equiv \frac{1}{2}\frac{\mathrm{d}\langle v^2 \rangle}{dt} = \langle \mathbf{v} \cdot \mathcal{F} \rangle - \nu \langle |\boldsymbol{\omega}|^2 \rangle \,, \quad (1.12)$$

where the mean square vorticity $\langle |\boldsymbol{\omega}|^2 \rangle$ is called the enstrophy. The second term on the right-hand-side is always negative. Thus in the absence of forcing the kinetic energy decreases monotonously due to viscous effects. In a forced stationary state the total energy is constant and there is a balance between the injection rate and the dissipation rate, which are the two terms in the right-hand-side of (1.12).

1.2　Laminar and turbulent flows

Depending on the forcing, boundary conditions and parameters, the Navier-Stokes equations describe a great variety of flows ranging from time-independent laminar flows with a simple structure to highly complex swirling turbulent flows. It was first observed by O. Reynolds that the complexity of the flow depends on a non-dimensional parameter – named after him as the Reynolds number – defined as the typical magnitude of the ratio of the terms representing inertia and viscosity in the momentum equation (1.8):

$$\mathrm{Re} = \frac{U(U/L)}{\nu U/L^2} = \frac{UL}{\nu}, \tag{1.13}$$

where L and U are characteristic length and velocity scales. Introducing the non-dimensional quantities $\mathbf{v}' = \mathbf{v}/U$, $t' = tU/L$, $\mathbf{x}' = \mathbf{x}/L$, $p' = p/(\rho U^2)$, $\mathcal{F}' = \mathcal{F}L/U^2$ (and dropping the primes) we obtain the non-dimensional form of the Navier-Stokes equation as

$$\frac{\partial \mathbf{v}}{\partial t} + \mathbf{v} \cdot \nabla \mathbf{v} = -\nabla p + \frac{1}{\mathrm{Re}} \nabla^2 \mathbf{v} + \mathcal{F}, \tag{1.14}$$

Reynolds performed experiments using a transparent tube with a fine streak of dye injected into the flow for visualization (Reynolds, 1883). For low Re (i.e. small velocity, thin tube or high viscosity) the streak of dye was straight indicating a simple time-independent unidirectional flow parallel to the central axis of the tube. For large Re (e.g. obtained by applying a higher pressure difference that produces larger flow velocities) the dye streak broke up into an irregular fluctuating pattern as a consequence of the complex whirling fluid motion.

Similar laminar-turbulent transitions for increasing Re can be observed in range of different types of flows. One well known example is the flow in the wake of an obstacle (e.g. a cylinder with its axis perpendicular to the mean flow). For low Reynolds numbers the fluid simply passes around the cylinder. Then, at slightly higher Re a steady recirculation zone is created behind the cylinder breaking the symmetry between the upstream and downstream regions. When the Reynolds number is increased further the flow in the recirculation zone becomes time-dependent. First the velocity field oscillates periodically in time, producing the so called Kármán vortex street. Then the flow becomes aperiodic and for high Re it has a completely irregular fluctuating structure both in space and time. Laminar-turbulent transition has also been studied extensively in fluid convection induced by a vertical temperature gradient producing an unstable density stratification (i.e. higher temperature at the lower boundary). Depending on the material properties of the fluid and the magnitude of the imposed temperature gradient there is a range of possible flow regimes including heat conduction with no flow, formation of a regular array of convection cells (rolls, hexagons etc.) with a time-independent velocity field, unsteady e.g. oscillating convection cells, and turbulent flow with an irregularly fluctuating structure in both space and time.

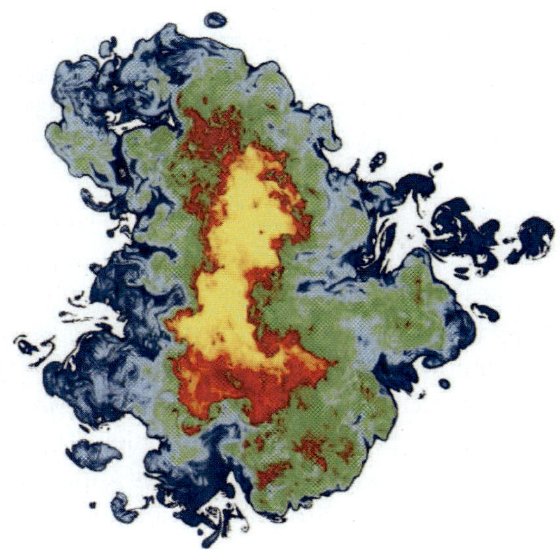

Figure 1.1: Complex multiscale flow structure shown by the concentration of a fluorescent dye in the cross section of a turbulent jet at Re $\sim$ 20000 (experiment by Catrakis et al. (2002)).

Note that the laminar-turbulent transition is not a sharp transition that takes place at a fixed value of some parameters but rather there is a range of Reynolds numbers in which simpler flows became unstable and a sequence of transitions (bifurcations) gradually leads to the formation of more and more complex flow structures on a broad range of length and time scales that is characteristic to fully developed turbulence (see Fig. 1.1). The details of the initial transitions and the corresponding values of Re depend on the type of flow considered, but typically it takes place around Re $\sim 10^3$.

Very large and very small Reynolds numbers both occur in real life flows. The flow around a bacteria of size about 1μm swimming in water at a typical speed of 15μm/s has a Reynolds number of about 10^{-5}, while for the airflow around a car ($L \approx 1$m) moving with a velocity $U = 100$ km/h, Re $\sim 10^6$. Large Reynolds number is also a characteristic feature of the large-scale geophysical flows with the well known consequence of the chaotic unpredictable nature of the daily weather.

When the Reynolds number is very small the inertial term can be neglected in the Navier-Stokes equation (1.14) and the velocity field is typically time-independent so that in the absence of external forces ($\mathcal{F} = 0$) the pressure differences are balanced by viscous forces

$$\nu\nabla^2\mathbf{v} = \frac{1}{\rho}\nabla p. \tag{1.15}$$

This is the Stokes equation that has analytic solutions for certain types of simple laminar flows, for example the laminar flow in a straight channel known as Poiseuille flow (Fig. 1.2). Assuming that the flow is unidirectional along the x axis ($v_y = v_z = 0$) it follows from incompressibility that the velocity profile $v_x(y)$ is uniform along the channel. When a pressure difference δp is applied along a channel of length L and width d the Stokes equation reduces to

$$\nu v_x''(y) = \frac{1}{\rho}\delta p L \ . \tag{1.16}$$

The solution consistent with the no-slip boundary conditions $v_x(y = \pm d/2) = 0$ has a parabolic velocity profile (Fig. 1.2) of the form

$$v_x(y) = \frac{\delta p}{2\nu\rho L}\left(\frac{d^2}{4} - y^2\right) \ . \tag{1.17}$$

A similar solution can be obtained for the flow in a cylindrical pipe where y is replaced by the radial distance from the axis of the cylinder. There are a few other simple analytic solutions of the Stokes equation, e.g. for the flow around a sphere, etc. (Lamb, 1932).

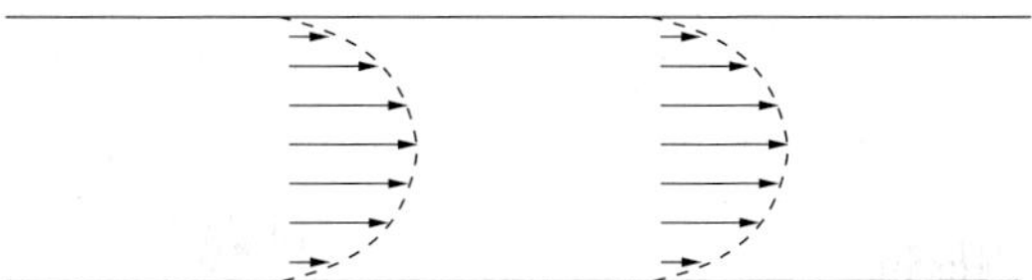

Figure 1.2: One of the simplest type of laminar flows: shear flow with a parabolic velocity profile in a straight channel.

Note that in fact the plane Poiseuille flow (1.17) is also an exact solution of the full Navier-Stokes equation. However, it was shown by linear stability analysis that this becomes unstable to small perturbations at a critical Reynolds number of 5772. In fact, the transition to turbulence is observed experimentally at even lower values of Re around 1000.

1.3 Turbulence

As already mentioned in the previous section, flows with a large Reynolds number have a strongly irregular, turbulent, character. Unfortunately it is not possible to obtain general analytic solutions of the Navier-Stokes equations for turbulent flows. Numerical solutions are also limited to moderate Reynolds numbers. This is due to the multiscale nature of the turbulent flows. As Re increases the range of length and time scales represented in the turbulent flow broadens and obtaining a good numerical approximation of the solution requires high resolution in space and time.

Another characteristic of turbulent flows is unpredictability, that is the high sensitivity of the solution to very small perturbations that are always present in real physical systems or numerical simulations. This unpredictability, also known as dynamical chaos, is a well known feature of much simpler low-dimensional nonlinear dynamical systems. Although in a strict mathematical sense a unique solution of the Navier-Stokes equation always exists for well-posed initial conditions (at least for large finite times), in practice the details of the forcing and boundary conditions are only known within some approximations and thus the solution in the turbulent regime represents only one of the possible typical realizations of the flow which are consistent with the initial and boundary conditions specified up to some finite precision. Although detailed information about a particular realization of a turbulent flow could be useful sometimes, for example in weather prediction, it is unnecessary in most situations. Therefore the general aim is to describe the statistical properties of the typical solutions such as mean values, typical deviations, probability densities etc. which are robust features of the flow and not affected by small perturbations.

An equation for the mean flow, defined as an average over many realizations of the turbulent velocity field, can be obtained (Reynolds, 1895) by decomposing the velocity and pressure fields into the sum of an ensemble-averaged and a fluctuating component:

$$\mathbf{v}(\mathbf{x},t) = \mathbf{V}(\mathbf{x},t) + \mathbf{v}'(\mathbf{x},t), \;\; p(\mathbf{x},t) = P(\mathbf{x},t) + p'(\mathbf{x},t) \qquad (1.18)$$

where $\mathbf{V} = \langle \mathbf{v} \rangle$ and $\mathbf{P} = \langle p \rangle$. The brackets $\langle ... \rangle$ denote now ensemble averages. Note that in the turbulent flow regime the fluctuations are not small compared to the mean flow. From the continuity equation it follows, assuming an incompressible fluid, that the mean flow and the fluctuations are both divergence free. Substituting (1.18) into the Navier-Stokes equation and averaging gives

$$\frac{\partial V_i}{\partial t} + \sum_{j=1,2,3} V_j \frac{\partial V_i}{\partial x_j} = -\frac{1}{\rho}\frac{\partial P}{\partial x_i} + F_i + \nu \frac{\partial^2 V_i}{\partial x_i^2} - \sum_{j=1,2,3} \frac{\partial \langle v_j' v_i' \rangle}{\partial x_j}. \qquad (1.19)$$

The averaging makes the fluctuations disappear from all the linear terms, but there is a contribution to the dynamics of the mean flow through the nonlinear term producing the last term on the right-hand side of (1.19). The symmetric tensor $\langle v_i' v_j' \rangle$ can be interpreted as an additional stress, known as the Reynolds stress, acting on the mean flow due to turbulence. The equation (1.19) for the mean flow is not closed. One can go further and obtain another equation for the correlations by subtracting (1.19) from the Navier-Stokes equation, but this will contain a term with third order correlations $\langle v_i' v_j' v_k' \rangle$, and so on for higher orders. This is known as the closure problem in turbulence, that is characteristic to nonlinear dynamical systems in general. The construction of approximate closure schemes for modelling turbulent flows has been an area of extensive research.

1.4 Kolmogorov's theory of turbulence

Richardson (1922) suggested the concept of an "energy cascade" for the description of turbulent flows. According to this the forcing injects energy into the flow and produces large eddies that became unstable and break up transferring their kinetic energy to somewhat smaller eddies. This "cascade" process generating smaller and smaller eddies continues until a smallest scale is reached at which the viscous forces dominate and the kinetic energy is converted into heat.

Significant progress towards a theoretical description of this cascade in homogeneous three-dimensional turbulent flows was made by Kolmogorov in a series a papers (Kolmogorov, 1941a,b). His theory is based on the assumption that at scales much smaller than the forcing scale (L) the turbulent flow has a universal statistically homogeneous and isotropic structure independent of the details of the forcing and of the large scale flow. Assuming statistical self-similarity of the velocity field the statistical properties of the flow describing the distribution of velocity fluctuations over different length scales can be expressed as universal functions of the energy dissipation rate and viscosity.

Kolmogorov further assumed that at sufficiently large Reynolds number and intermediate length scales, that are much smaller than the forcing scale but larger than the length scale where viscous dissipation becomes important, the statistics of the velocity field is independent of the viscosity and only depends on the overall energy dissipation rate. A consequence of this is that the energy dissipation rate per unit mass must tend to a non-zero constant in the limit of vanishing viscosity:

$$\lim_{\nu \to 0} \nu \langle (\nabla \times \mathbf{v})^2 \rangle = \epsilon > 0 \,. \tag{1.20}$$

This appears to be in agreement with experimental observations. Since the viscosity term has the highest derivative in the Navier-Stokes equation the limit $\nu \to 0$ is singular and corresponds to divergent velocity gradients. This is an example of a dissipative anomaly in which the time-reversal symmetry, that is broken for the viscous flow, is not restored in the limit of vanishing viscosity.

Using the dissipation rate (ϵ) as the only relevant flow parameter in the $\nu \to 0$ limit, from dimensional considerations the typical fluctuations of the velocity field over a distance l and the corresponding characteristic timescales $\tau(l)$, usually interpreted as "eddy turnover time", can be estimated as

$$\delta v(l) \sim (\epsilon l)^{1/3}, \quad \tau(l) = \frac{l}{v(l)} \sim \epsilon^{-1/3} l^{2/3}. \tag{1.21}$$

Thus, in the Re $\to \infty$ limit the turbulent velocity field is a rough non-differentiable function $\delta v(l) \sim l^{1/3}$ for $l \to 0$. This expression identifies the so-called Hölder exponent as $\alpha = 1/3$. It is easy to see that the energy transfer rate $\delta v(l)^2 / \tau(l)$ is independent of the length scale l and applying the first of the above formulas to the integral scale (L) one can relate the energy dissipation rate to the large scale properties of the flow as $\epsilon \sim U^3/L$.

The results of this dimensional analysis are supported by an exact result obtained by Kolmogorov for the third-order longitudinal structure function. The structure functions are moments of velocity differences measured at a given distance. More specifically the structure function of order n, $S_n(l)$, is defined as

$$S_n(l) \equiv \langle [(\mathbf{v}(\mathbf{x}+\mathbf{l}) - \mathbf{v}(\mathbf{x})) \cdot \mathbf{l}/l]^n \rangle . \tag{1.22}$$

Brackets denote spatial or ensemble averages. Starting from the Navier-Stokes equation and assuming a statistically isotropic and homogeneous velocity field one can derive the Kármán-Howarth equation for the longitudinal structure functions as

$$\frac{\partial S_2}{\partial t} + \frac{1}{3l^4} \frac{\partial}{\partial l}(l^4 S_3) + \frac{4}{3}\epsilon = \frac{2\nu}{l^4} \frac{\partial}{\partial l}\left(l^4 \frac{\partial S_2}{\partial l}\right). \tag{1.23}$$

Taking the $\nu \to 0$ limit in a stationary state the third moment of the longitudinal velocity differences is obtained as

$$S_3(l) = -\frac{4}{5}\epsilon l. \tag{1.24}$$

Note that this is consistent with the scaling (1.21) based on the universality and self-similarity assumptions ($\delta v(l) = S_1(l)$).

Using the statistical self-similarity hypothesis this exact result can be generalized to the nth order structure functions as

$$S_n(l) = C_n \epsilon^{n/3} l^{n/3}, \tag{1.25}$$

where C_n are dimensionless constants.

Alternatively, the fluctuating velocity field can be characterized by the energy spectrum defined as the Fourier transform of the two-point velocity correlation integrated over a spherical shell of wavenumbers with magnitude k:

$$E(k) = \frac{1}{2(2\pi)^3} \int \left[\int \langle \mathbf{v}(\mathbf{x}) \cdot \mathbf{v}(\mathbf{x} + \mathbf{r}) \rangle e^{-i\mathbf{kr}} \mathrm{d}\mathbf{r} \right] \delta(|\mathbf{k}| - k) \mathrm{d}\mathbf{k}. \tag{1.26}$$

From Kolmogorov's hypotheses it follows that the energy spectrum at large Re has a form $E(k) = f(k, \epsilon)$ and dimensional analysis yields

$$E(k) = C_K \epsilon^{2/3} k^{-5/3} \tag{1.27}$$

where C_K is called the Kolmogorov constant that from experimental data is around $C_K \simeq 1.5$. The spectral exponent $\gamma = 5/3$ can also been obtained from the mathematical relationship between γ and the exponent of the second-order structure function $S_2 \sim l^{\zeta_2}$, as $\gamma = 1 + \zeta_2$.

For large, but finite Reynolds numbers the above scalings remain valid in the range of lengthscales where both the energy injection due to external forces and the viscous dissipation can be neglected. Viscous dissipation becomes important at the lengthscale where the eddy turnover time $\tau(l)$ is comparable to the characteristic time of diffusion of momentum l^2/ν. Alternatively one can define a scale-dependent effective Reynolds number that measures the ratio of the inertial and viscous terms for eddies of size l as $\mathrm{Re}(l) = \delta v(l) l / \nu$. For turbulent flows with $\mathrm{Re} = \mathrm{Re}(L) \gg 1$ there is a range of scales where viscous dissipation is negligible and the kinetic energy flows to smaller scales without any losses. The lengthscale η where this energy cascade ends is thus given by the condition

$$\mathrm{Re}(\eta) = \frac{\delta v(\eta) \eta}{\nu} \simeq \mathcal{O}(1). \tag{1.28}$$

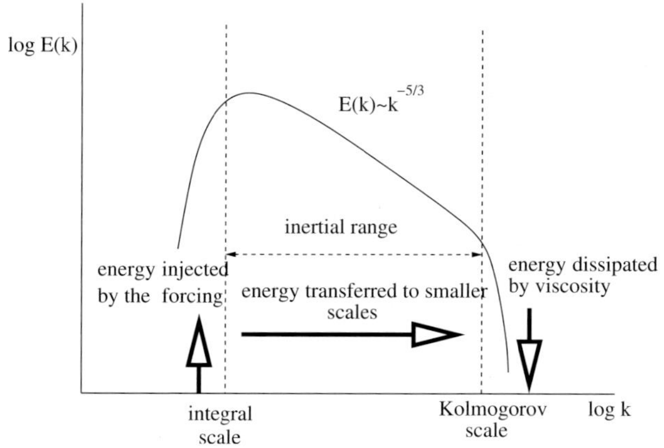

Figure 1.3: Schematic diagram of the kinetic energy spectrum of a turbulent flow.

Using (1.21) this gives

$$\eta \sim \left(\frac{\nu^3}{\epsilon}\right)^{\frac{1}{4}} \sim \left(\frac{L\nu^3}{U^3}\right)^{\frac{1}{4}} = \frac{L}{\mathrm{Re}^{3/4}}. \qquad (1.29)$$

This is the Kolmogorov scale that represents the size of the smallest flow structures in the turbulent velocity field. Thus, the range of validity of the scalings in (1.21) and (1.25) is given by the so called inertial range where $L \gg l \gg \eta$ (see Fig. 1.3). Note that the lengthscale ratio of the largest and smallest eddies increases with the Reynolds number as $\mathrm{Re}^{3/4}$. Thus the number of gridpoints or Fourier modes needed to fully resolve a three-dimensional turbulent flow in a numerical simulation can be estimated as $(L/\eta)^3 \sim \mathrm{Re}^{9/4}$ indicating that many of the naturally occurring highly turbulent flows are well beyond the reach of the capacity of current computers.

Experiments are in good agreement with the predicted $-5/3$ energy spectrum (1.27), but there are significant deviations from (1.25) for higher order structure functions with $n > 3$. They follow power laws $S_n(l) \sim l^{\zeta_n}$, but with scaling exponents $\zeta_n < n/3$ that do not satisfy the self-similarity hypotheses (Sreenivasan and Antonia, 1997). These deviations are due to the intermittent, strongly non-homogeneous character of the velocity fluctuations, that cannot be described by a single constant value of the dissipation rate. Currently, there is no satisfactory theoretical description for the experimentally measured anomalous scaling exponents. Further information on Kolmogorov's theory, their corrections, and turbulent flows in general, can be found in books such as Batchelor (1953), Frisch (1995), Lesieur (1990), or Pope (2000).

1.5 Two-dimensional flows

Although strictly speaking there are no two-dimensional flows in nature there has been considerable research on turbulent flows in two spatial dimensions (see the reviews by Kraichnan and Montgomery (1980) or Tabeling (2002)). This has been partly motivated by the interest in geophysical flows that can be considered quasi-two-dimensional since stable density stratification and the Earth's rotation both restrict large scale motions to two-dimensional layers. Quasi-two-dimensional turbulent flows have also been studied in laboratory experiments with soap films, density stratification or using rotating tanks. Obviously, two-dimensional turbulence is also more accessible for high resolution numerical simulations.

It turns out that there are significant differences in some aspects of turbulence between three- and two-dimensional systems. As the vorticity vector points in the direction perpendicular to the plane of the flow $\boldsymbol{\omega} \perp \mathbf{v}$, fluid motion can be fully described by a scalar field $\omega(x, y)$. A consequence of two-dimensionality is that the vortex stretching term $\boldsymbol{\omega} \cdot \nabla \mathbf{v}$ vanishes in the vorticity equation (1.11) that becomes

$$\frac{\partial \omega}{\partial t} + \mathbf{v} \cdot \nabla \omega = \nu \nabla^2 \omega + |\nabla \times \mathcal{F}|. \tag{1.30}$$

Thus, in the absence of the forcing and viscous dissipation the vorticity is conserved along the trajectories of fluid elements. Equation (1.30) describes the evolution of the vorticity distribution in a given velocity field. However, ω and $\mathbf{v}$ are obviously not independent of each other, but are two alternative representations of the instantaneous flow field. The velocity field is a vector, but it is subjected to the incompressibility condition. This constraint can be eliminated in two-dimensional flows by introducing the scalar streamfunction $\psi(x, y, t)$ that defines the flow as

$$v_x = -\frac{\partial \psi}{\partial y}, \quad v_y = \frac{\partial \psi}{\partial x} \tag{1.31}$$

and automatically satisfies incompressibility. Taking the curl of the velocity field one obtains a relationship between the streamfunction and vorticity field as

$$\omega(x, y) = \frac{\partial^2 \psi}{\partial x^2} + \frac{\partial^2 \psi}{\partial y^2}. \tag{1.32}$$

Thus the dynamics can be described from (1.30) as the vorticity field advected by the flow defined by the streamfunction (1.31) while the new streamfunction at a later time can be obtained from the vorticity field by inverting (1.32). This second step is computationally expensive since it requires the solution of an elliptic equation in which the streamfunction at any given point depends on the whole distribution of vorticity.

In the case of a two-dimensional inviscid fluid with no forcing a special class of simple analytic solutions is a system of point vortices represented as a singular distribution of vorticity concentrated on a set of points

$$\omega(\mathbf{x}, t) = \sum_i \Gamma_i \delta(|\mathbf{x} - \mathbf{r}_i(t)|), \tag{1.33}$$

where Γ_i and $\mathbf{r}_i$ are the strength and position of the ith vortex, respectively. The streamfunction for the velocity field generated by the point vortex system can be obtained from (1.32) as the Green's function of the Laplace equation

$$\psi(\mathbf{x}, t) = -\sum_i \frac{\Gamma_i}{2\pi} \log |\mathbf{x} - \mathbf{r}_i(t)| \tag{1.34}$$

and the velocity components of a single point vortex in polar coordinates with the origin at the vortex center are

$$v_r = 0, \quad v_\phi = \frac{\Gamma}{2\pi r}. \tag{1.35}$$

Since the vorticity is conserved within the fluid elements the equation of motion for the vortex centers in a system of point vortices is given by the advection equation $\dot{\mathbf{r}}_i = \mathbf{v}[\mathbf{x} = \mathbf{r}_i(t), t]$. Each point vortex is passively advected just like any other fluid element by the superposition of the velocity fields generated by all the other vortices (excluding self-interaction). Thus the equation of motion of a system of point vortices is given by

$$\dot{x}_i = -\frac{1}{2\pi} \sum_{j \neq i} \frac{\Gamma_j(y_i - y_j)}{r_{ij}^2}, \quad \dot{y}_i = \frac{1}{2\pi} \sum_{j \neq i} \frac{\Gamma_j(x_i - x_j)}{r_{ij}^2}, \qquad (1.36)$$

where $r_{ij} = |\mathbf{r}_i - \mathbf{r}_j|$ is the distance between vortices i and j. This type of point vortex systems exhibit a great variety of flows (Aref, 1983) and have been used as model flows in theoretical studies of mixing.

More complex turbulent-flow solutions appear in the presence of forcing and non-zero viscosity. As mentioned earlier the vortex stretching mechanism for transferring energy to smaller scales does not operate in two-dimensional turbulent flows. In the absence of forcing and viscosity the vorticity is conserved along the trajectories, $D\omega/Dt = 0$. Thus vorticity can not diverge in the inviscid limit and the viscous energy dissipation rate $\nu\langle\omega^2\rangle$ vanishes as $\nu \to 0$. Thus, from (1.12) the energy cannot be removed by viscosity at small scales in the $\nu \to 0$ limit. This rules out the possibility of the direct energy cascade to smaller scales as in three-dimensional turbulence. Kraichnan (1967) suggested that the energy injected into the flow by the forcing is transferred to larger scales resulting in an "inverse energy cascade". In a finite system the energy is eventually dissipated by friction acting at the boundaries or by other large scale processes. In numerical simulations a stationary state can be established by introducing some sort of damping to prevent the accumulation of energy at the largest scales set by the size of the computational domain. Dimensional arguments imply that the energy spectrum for the inverse cascade range, i.e. between the largest scale and the forcing scale has the same form as in three dimensions: $E(k) = C'\epsilon^{2/3}k^{-5/3}$ and the predicted $k^{-5/3}$ power law has been verified in experiments and numerical simulations (Tabeling, 2002). The scaling of the structure functions follows $S_n(l) \sim l^{n/3}$ and in contrast with three-dimensional turbulence the experimental data do not show significant deviations from this due to intermittency, in this large scale range.

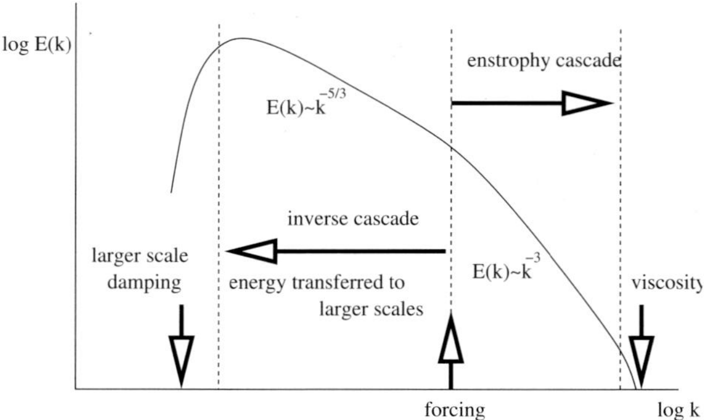

Figure 1.4: Schematics of the dual cascade energy spectrum in two-dimensional turbulence.

Batchelor (1969) and Kraichnan (1967) also proposed the existence of another cascade in two-dimensional flows that is associated with the transfer of the enstrophy to smaller scales through the formation of thin vorticity filaments due to the large scale strain generating strong vorticity gradients. The enstrophy cascade ends at a small length scale where enstrophy is dissipated by the diffusion of vorticity due to molecular viscosity. Assuming a constant flux of enstrophy gives an energy spectrum of the form $E(k) \sim \theta^{2/3} k^{-3}$ at scales below the forcing scale, where $\theta = \nu \langle (\nabla \omega)^2 \rangle$ is the enstrophy dissipation rate. Similarly to the energy dissipation rate in three-dimensional turbulence the enstrophy dissipation rate converges to a finite constant in the limit of vanishing viscosity. At finite viscosity the enstrophy cascade ends at the length scale $l_Z = \theta^{-1/6} \nu^{1/2}$.

Note that in contrast to the "rough" velocity field of three-dimensional turbulence, the k^{-3} spectrum implies an almost everywhere smooth velocity field at small scales such that $\delta v(l) \sim l$ (with logarithmic corrections). This also means that at scales below the forcing scale, within the enstrophy cascade range, the flow has a single characteristic timescale: $\tau(l) \sim l/\delta v(l) \approx \tau^*$, which is independent of the lengthscale l.

The theoretically predicted dual cascade with two power-law regimes in the kinetic energy spectrum (Fig. 1.4) has been reproduced in numerical simulations and confirmed by laboratory experiments. In some of the experiments spectra steeper than k^{-3} was observed in the enstrophy cascade range. This deviation can be related to the presence of additional damping at large scales, the so-called Ekman friction. Since the theoretical description of this regime is very similar to the problem of chemical decay in smooth flows we will return to this later in Chapter 6.

Chapter 2

Mixing and Dispersion in Fluid Flows

2.1 Introduction

Physical, chemical or other type of properties of a fluid medium –
temperature, concentration of pollutants, density of suspended particles or microorganisms – are often distributed non-uniformly in
space. In a moving fluid the flow modifies the spatial distributions
by transporting and blending together fluid masses of different properties. This plays an important role in a range of natural and tech-
nological processes including large scale geophysical flows, chemical
reactors, microfluidic devices, etc.

In a closed system with a flow on a finite domain the transport
processes tend to reduce the inhomogeneities gradually producing
a more and more uniform distribution. However, even when the
asymptotic state is completely uniform often this is only reached after long time and the characteristics of the transient behavior can
strongly depend on the properties of the flow. In the presence of
non-uniformly distributed sources and sinks of the transported quantities, inside the domain or along its boundaries, the spatial non-
uniformities persist indefinitely resulting in a statistically stationary
state in which there is a balance between the creation of new inhomogeneities by the sources and the homogenization effect of mixing.

The aim of a theoretical description of mixing is to characterize the spatial distribution of transported quantities in a stationary state and to describe the transient process of homogenization in various types of flows, relating these to relevant characteristics of the velocity field. Even when the velocity field is given explicitly (e.g. as an analytical model flow or in the form of data from observations or simulation) deducing its consequences for transport can be highly non-trivial. It turns out that even simple flows can create very complicated strongly non-uniform distribution of the transported quantities. This is further complicated by the wide range of qualitatively different types of flows from simple laminar to complex turbulent flows that can only be specified through some statistical properties.

An important intermediate level of description of mixing can be given in terms of the trajectories of fluid elements in the flow. This is the so-called Lagrangian description. Various characteristics of the ensemble of trajectories, like absolute and relative dispersion, contain useful information for predicting the evolution of the spatial distribution of quantities of interest.

Although in certain circumstances the distribution of some transported properties of the fluid may influence the velocity field itself (e.g. temperature distribution in thermal convection, density of bacteria in case of bioconvection, exothermic chemical reactions in combustion) in the following we will only consider the mixing of *passive* components, i.e. we assume that the fluid dynamics process generating the underlying velocity field is independent of the transport and mixing so the latter can be treated separately as a dynamics superimposed on a prescribed flow. This simplifying assumption is valid for many problems of practical interest. In this chapter we focus on mixing of *inert* substances transported by a flow that may have sources or sinks, but do not participate in processes like chemical reactions, biological interactions etc. Those will be discussed in the following chapters.

2.1.1 Advection

The dominant transport process in a moving fluid is advection that consists of the rearrangement of fluid elements in space as a result of the motion of the medium. The distribution of suspended particles or of some properties of the fluid can be characterized by a continuous density or concentration field $C(\mathbf{x}, t)$. The evolution of the concentration of a passively advected inert substance is governed by the general conservation equation (1.1) for C:

$$\frac{\partial C}{\partial t} + \nabla \cdot \mathbf{J} = S(\mathbf{x}, t) \, , \qquad (2.1)$$

where $S(\mathbf{x}, t)$ specifies the distribution of sources and sinks of this substance. When the flux $\mathbf{J}$ is just given by the advection of fluid particles by the velocity field, it is proportional to the local velocity $\mathbf{J_A} = \mathbf{v}C$. In this case, and particularizing to an incompressible flow, Eq. (2.1) becomes

$$\frac{\partial C}{\partial t} + \mathbf{v} \cdot \nabla C = S(\mathbf{x}, t) \, . \qquad (2.2)$$

The left hand side is the Lagrangian derivative of C, that is the rate of change of the concentration along a path following a fluid element. This allows to write the time evolution of the concentration C in a fluid element as:

$$\frac{D}{Dt} C(\mathbf{x}, t) = \frac{dC(t)}{dt} = S(t), \qquad (2.3)$$

where $C(t) \equiv C(\mathbf{x} = \mathbf{r}(t), t)$ and $S(t) \equiv S(\mathbf{x} = \mathbf{r}(t), t)$ are the concentration at a particular fluid element and the rate at which the substance is injected into it along its path $\mathbf{r}(t)$, which satisfies

$$\dot{\mathbf{r}} = \mathbf{v}[\mathbf{r}(t), t]. \qquad (2.4)$$

This equation also describes the motion of very small suspended particles with negligible inertia that take on the local velocity of the medium instantaneously. The ensemble of solutions corresponding to all the initial positions within the domain of the flow $\mathbf{r}_0 \in \mathcal{D}$ defines a mapping between initial and final positions of each fluid element

$$\mathbf{r}(t) = \mathbf{\Phi}_{0,t}(\mathbf{r_0}). \qquad (2.5)$$

The Lagrangian map $\mathbf{\Phi}$ gives a full description of the flow that is equivalent to the so called Eulerian description based on specifying the velocity field $\mathbf{v}(\mathbf{x}, t)$ relative to a fixed reference frame. When no confusion on the initial time could arise, we will use the notation $\boldsymbol{\Phi}_t$ instead of $\boldsymbol{\Phi}_{0,t}$.

From the Lagrangian map, the velocity field can be easily recovered by differentiation with respect to time. The inverse procedure, obtaining the Lagrangian map from the velocity field, requires integration of (2.4) and it is usually not possible doing it other than numerically. In any case, one can express the formal solution of equation (2.2) in terms of the solution of (2.3) and the inverse of the Lagrangian map $\boldsymbol{\Phi}^{-1}$ as

$$C(\mathbf{x}, t) = C[\boldsymbol{\Phi}_{0,t}^{-1}(\mathbf{x}), t = 0] + \int_0^t S[\boldsymbol{\Phi}_{t',t}^{-1}(\mathbf{x}), t']dt' \qquad (2.6)$$

that gives the concentration at any given point as the sum of the initial concentration at the original position of the corresponding fluid element and the contribution of sources and sinks accumulated along its trajectory.

In the absence of sources, pure advection conserves the concentrations within fluid elements and only rearranges their distribution in space leaving the probability density function of the concentrations unchanged. Therefore to capture the gradual homogenization of an initially non-uniform concentration field under mixing a second transport process – diffusion – also needs to be included in the description.

2.1.2   Diffusion

Diffusion represents the transport due to the irregular thermal motion at molecular scales (other types of irregular motion may also be modelled by diffusion as discussed later). While advection conserves the content of fluid elements, diffusion is the process in which fluid elements interchange contents when they are sufficiently close to each other. In the framework of continuum description this is an essential step by which chemical molecules initially introduced in a system at different locations may come together at the same fluid element and thus really react.

The effect of transport by molecular diffusion on a concentration field $C(\mathbf{x}, t)$ can be obtained from the conservation equation

(Eq. (2.1)) under the assumption that the diffusive flux is proportional to the local concentration gradient (Fick's law)

$$\mathbf{J}_D = -D\nabla C \tag{2.7}$$

that leads, in the absence of sources, to the diffusion equation

$$\frac{\partial C}{\partial t} = \nabla \cdot (D\nabla C) . \tag{2.8}$$

The proportionality factor D is the diffusion coefficient which in the simplest cases is assumed to be a constant independent of the concentration and uniform in space. In this case

$$\frac{\partial C}{\partial t} = D\nabla^2 C. \tag{2.9}$$

When molecular motion is the only mechanism contributing to diffusion, D is proportional to the temperature and is given by the Einstein relation $D = kT/\gamma$, where k is the Boltzmann constant and γ is the drag coefficient in the fluid.

The general solution of (2.9) can be written as

$$C(\mathbf{x},t) = \int d\mathbf{x}' G(\mathbf{x} - \mathbf{x}',t)C_0(\mathbf{x}') , \tag{2.10}$$

where $C_0(\mathbf{x}) = C(\mathbf{x}, t = 0)$ and $G(\mathbf{x},t)$ is the fundamental solution of (2.9) or Green function, i.e. the solution with initial condition $C_0(\mathbf{x}) = \delta(\mathbf{x})$. For boundary conditions such that $C(\mathbf{x},t) \to 0$ when $|\mathbf{x}| \to \infty$ this solution is

$$G(\mathbf{x},t) = \frac{1}{(4\pi Dt)^{d/2}} e^{-\frac{\mathbf{x}^2}{4Dt}}. \tag{2.11}$$

This shows that the result of the diffusion process starting from a localized concentration or density patch is a growing isotropic concentration cloud whose width, defined as

$$w^2 \equiv \frac{\int \mathbf{x}^2 G(\mathbf{x},t)d\mathbf{x}}{\int G(\mathbf{x},t)d\mathbf{x}} , \tag{2.12}$$

grows like

$$w \approx (2dDt)^{1/2} , \tag{2.13}$$

where d is the number of spatial dimensions. Typical diffusion coefficients of solute molecules in liquids are $D \approx 10^{-9}$ m^2/s. This means that about 10 minutes are needed for the molecular diffusion to disperse concentrations over a distance of the order of one millimeter and half a day to disperse to centimeter scales. The time becomes half a millisecond for distances of the order of 1 μm. Thus molecular diffusion alone is a very inefficient process at any macroscopic scale, although it plays an important role at biological cell dimensions and below, down to molecular scales.

Although the typical concentration isolines expand as $w \sim t^{1/2}$ as time increases, (2.10) and (2.11) also predict that for any initial condition a non-vanishing concentration can be found arbitrarily far from its initial support as soon as $t > 0$. This is clearly an unphysical feature of the continuum description that looses its validity where the concentration is too small. In this case a description in terms of particles that can not travel arbitrarily far in a very short time is more appropriate.

The diffusion equation can also be derived as a continuum limit for the equation governing the distribution of a large set of independent particles performing random walks, composed of a sequence of uncorrelated random steps. In addition to molecules, any uncorrelated motion of a large number of objects unavoidably leads, as a consequence of the central limit theorem, to a dispersion that follows the mathematical rules of molecular diffusion, although now the diffusion coefficient will not be related to thermal motion and can be much larger. This idea is usually used to model dispersion of biological populations in terms of diffusion equations (Okubo and Levin, 2002) with a diffusion coefficient larger than molecular values. For example, in the case of a one-dimensional random walk with jumps separated by time τ and step-size distribution $P_\tau(\delta)$ the probability density for the position of a walker (which is equal to the expected density distribution of a large set of independent walkers) at time $t + \tau$ satisfies

$$\rho(x, t + \tau) = \int \rho(x - \delta, t) P_\tau(\delta) d\delta. \tag{2.14}$$

Considering the distribution of particles at scales much larger than the typical values of the individual jumps the density inside the integral can be Taylor expanded with respect to δ

$$\rho(x, t+\tau) = \rho(x, t) \int P_\tau(\delta) - \frac{\partial \rho}{\partial x} \int \delta P_\tau(\delta) d\delta$$
$$+\frac{1}{2}\frac{\partial^2 \rho}{\partial x^2} \int \delta^2 P_\tau(\delta) d\delta + \cdots . \qquad (2.15)$$

For an unbiased symmetric random walk, $P(\delta) = P(-\delta)$, the second term on the right vanishes and taking the time-continuous limit of small τ one obtains a diffusion equation with diffusion coefficient determined by the variance of the jumps: $D = \langle \delta_\tau^2 \rangle / (2\tau)$. In d dimensions the result is $D = \langle \boldsymbol{\delta}_\tau^2 \rangle / (2d\tau)$. In the random walk context, the dispersion in Eq. (2.13) is giving the second moment of the position of a random walker which started at $\mathbf{r} = 0$:

$$\langle \mathbf{r}^2 \rangle = w^2 = \int \rho(\mathbf{r}, t) \mathbf{r}^2 d\mathbf{r} = 2dDt , \qquad (2.16)$$

with $\rho(\mathbf{r}, t=0) = \delta(\mathbf{r})$.

Turbulent diffusion

As shown by Taylor (1921), dispersion produced by advection in an irregular turbulent flow can also be described as a diffusion process. The starting element is the position of a particle initially at $\mathbf{r}_0$, as obtained by integrating Eq. (2.4):

$$\mathbf{r}(t) = \mathbf{r}_0 + \int_0^t \mathbf{v}[\mathbf{r}(t'), t']dt'. \qquad (2.17)$$

Assuming that the velocity field is a random function of space and time, the expectation value of the particles final position can be obtained by averaging the above equation over many realizations of the random velocity field. If the mean velocity is zero this gives the simple result $\langle \mathbf{r}(t) \rangle = \mathbf{r}_0$. The standard deviation of the distance from the initial point satisfies the equation

$$\frac{d}{dt}\langle (\mathbf{r} - \mathbf{r}_0)^2 \rangle = 2\langle (\mathbf{r} - \mathbf{r}_0) \cdot \mathbf{v}[\mathbf{r}(t)] \rangle. \qquad (2.18)$$

Integrating both sides gives

$$\langle (\mathbf{r} - \mathbf{r}_0)^2 \rangle = 2 \int_0^t dt' \int_0^{t'} dt'' \langle \mathbf{v}[\mathbf{r}(t')] \cdot \mathbf{v}[\mathbf{r}(t' - t'')] \rangle \; , \qquad (2.19)$$

where inside the integrals we have a Lagrangian velocity correlation function

$$f_L(t', t'') \equiv \frac{1}{\langle v^2 \rangle} \langle \mathbf{v}[\mathbf{r}(t')] \cdot \mathbf{v}[\mathbf{r}(t' - t'')] \rangle \qquad (2.20)$$

which is the correlation between velocities at different times along the trajectory of a fluid element. If the turbulent flow is stationary and homogeneous this correlation function should only depend on the time difference $\tau = |t'' - t'|$

$$\langle (\mathbf{r} - \mathbf{r}_0)^2 \rangle = 2 \langle v^2 \rangle \int_0^t dt' \int_0^{t'} dt'' f_L(|t'' - t'|) \; , \qquad (2.21)$$

where $f_L(0) = 1$. A typical *Lagrangian correlation time*, T_L could be defined as

$$T_L \equiv \int_0^{\infty} d\tau f(\tau) \; . \qquad (2.22)$$

It would indicate the time after which the correlation $f_L(t)$ becomes negligible. If T_L is finite we can distinguish two regimes. For short times $t \ll T_L$ we can approximate $f_L(\tau) \sim 1$ that gives a ballistic behavior

$$\langle (\mathbf{r} - \mathbf{r}_0)^2 \rangle \simeq \langle v^2 \rangle t^2 \qquad (2.23)$$

in which displacement increases linearly with time. For long times, when $t \gg T_L$, the limit in the second integral can be extended to infinity since the velocities are uncorrelated for large times, giving

$$\langle (\mathbf{r} - \mathbf{r}_0)^2 \rangle = \langle v^2 \rangle T_L t \qquad (2.24)$$

In this regime the typical distance from the origin of motion increases as the square root of time. Thus, the dispersion in turbulent flows at long times is analogous to molecular diffusion or random walks with independent increments and comparison of Eq. (2.24) with (2.16) relates the turbulent diffusion coefficient, D_T, to the integral of the Lagrangian correlation function, T_L, as

$$D_T \simeq \frac{1}{2d} \langle v^2 \rangle T_L. \qquad (2.25)$$

Anomalous diffusion

In some types of flows the dispersion of fluid elements exhibits an *anomalous diffusion* (Metzler and Klafter, 2000; Klages et al., 2008), where the width of the dispersing cloud of particles increases as

$$w \sim t^{\alpha} \ . \tag{2.26}$$

$\alpha > 1/2$ indicates superdiffusion and $\alpha < 1/2$ is the case of subdiffusion. This type of behavior arises when either $\langle v^2 \rangle$ or T_L on the right-hand-side of (2.25) do not have well defined positive finite values. This could be due to the presence of coherent structures like jets and vortices in the flow that may lead to anomalous diffusion as a result of the trapping of particles inside localized vortical regions combined with long flights along the jets. Such anomalous diffusion of particles advected in an unsteady flow formed by a chain of vortices surrounded by jet regions has been found experimentally by Solomon et al. (1993, 1994). Wu and Libchaber (2000) also observed anomalous dispersion of passive particles in a flow generated by swimming bacteria whose collective motion in a concentrated suspension can create coherent flow structures (Dombrowski et al., 2004).

Anomalous diffusion can be modelled by the random walk processes known as Lévy flights (Weeks et al., 1996; Metzler and Klafter, 2000) where the distribution of the step size δ decays slowly, i.e. has a *fat tail* of power law form $P(\delta) \sim |\delta|^{-\mu}$ for large δ (see Fig. 2.1). If the step distribution decays sufficiently fast, $\mu > 3$, the dispersion is normal diffusion with $\alpha = 1/2$. When $\mu < 3$ the dispersion is dominated by rare large jumps that have a relatively high probability and the variance of the step size is infinite. In this case the central limit theorem does not apply and the asymptotic behavior cannot be described by normal diffusion. The resulting superdiffusion exponent is $\alpha = 2 - \mu/2$ for $2 < \mu < 3$ and $\alpha = 1$ for $1 < \mu < 2$, which corresponds to ballistic motion, i.e. the size of the particle cloud increases proportionally with time, like particles that move away in different directions with constant speed on straight lines. This type of random walk models can be extended to include a distribution of waiting times between the steps, e.g. to model the trapping of particles in vortex regions. Assuming again a power law distribution at large waiting times $P(\tau) \sim \tau^{-\nu}$, when $\mu > 3$ this leads to subdiffusion if $1 < \nu < 2$ with $\alpha = (\nu-1)/2$. If the step-size distribution decays more slowly, $1 < \mu < 3$, then the dispersion exponent is $\alpha = 1 + (\nu - \mu)/2$ that can be either sub- or superdiffusive depending on whether the long waiting times or the long jumps dominate.

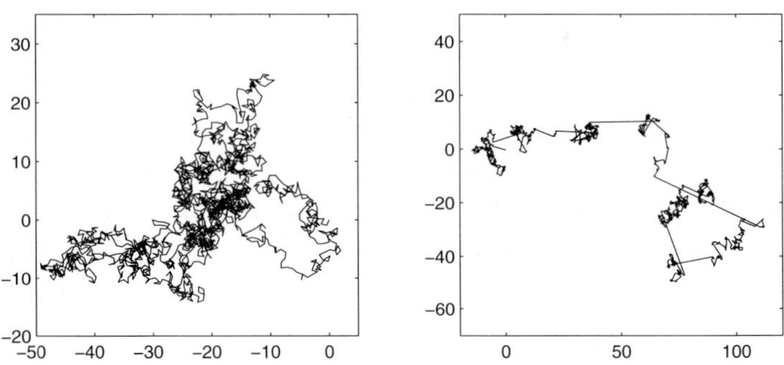

Figure 2.1: Normal diffusion in a two-dimensional random walk with a constant step size (left) and anomalous diffusion in a Lévy flight with step-size distribution $P(\delta) \sim \delta^{-5/2}$ (right).

Anomalous diffusion of a continuous concentration field can be modelled in terms of *fractional* differential equations. To see how they arise we can write Eq. (2.9) for normal diffusion in terms of the spatial Fourier transform of the concentration field $\hat{C}(\mathbf{k}, t)$. This can be easily done under periodic boundary conditions or in unbounded space as

$$\frac{\partial \hat{C}(\mathbf{k}, t)}{\partial t} = -D\mathbf{k}^2 \hat{C}(\mathbf{k}, t). \tag{2.27}$$

The solution of (2.27) is

$$\hat{C}(\mathbf{k}, t) = \hat{C}_0(\mathbf{k})e^{-Dk^2 t} \tag{2.28}$$

that under inverse Fourier transform reproduces (2.10) – (2.11). To obtain anomalous diffusion, Eq. (2.26), one can replace the exponent $Dk^2 t$ in (2.28) by $D|k|^\beta t$, or (2.27) by

$$\frac{\partial \hat{C}(\mathbf{k}, t)}{\partial t} = -D|\mathbf{k}|^\beta \hat{C}(\mathbf{k}, t) \tag{2.29}$$

with $\beta = 1/\alpha$. Symbolically this can be represented in real space by a *fractional derivative* equation:

$$\frac{\partial C(\mathbf{x}, t)}{\partial t} = D \nabla^\beta C(\mathbf{x}, t) \ . \tag{2.30}$$

We stress that all the approaches presented in this subsection, leading to normal and anomalous diffusion, are intended to model *dispersion*. As clearly seen in (2.28) and their equivalents for anomalous diffusion these equations damp the higher Fourier modes so that there is a relative increase of variance at larger scales. But in situations such as in modelling dispersion induced by turbulent flows the dispersion is a consequence of the stretching of fluid elements that is associated with compression along the complementary directions. This produces a direct cascade of variance towards small scales that is not captured by the diffusion equation (2.9). Thus the small scale distribution and phenomena strongly dependent on this, like chemical reactivity, may not be adequately modelled in the framework of diffusion equations alone.

2.1.3   Advection and diffusion

Taking into account both advective and diffusive transport, the evolution of the concentration of a passive non-reactive substance in an incompressible flow is governed by the advection-diffusion equation

$$\frac{\partial C}{\partial t} + \mathbf{v} \cdot \nabla C = S(\mathbf{x}) + D \nabla^2 C. \tag{2.31}$$

Stirring by advection brings together fluid parcels of different properties producing strong gradients in the concentration field that enhances mixing by molecular diffusion. The relative importance of the advective and diffusive transport on the characteristic length scale of the flow (L) is characterized by the Péclet number

$$\mathrm{Pe} = \frac{U/L}{D/L^2} = \frac{UL}{D} \ , \tag{2.32}$$

where U is the characteristic flow velocity. In most cases the Péclet number is large reflecting that transport by molecular diffusion is very slow on macroscopic length scales. For example, diffusive transport times for a solute like sugar or salt over a distance of a few centimeters in a liquid is longer than a day. However, efficient transport and homogenization in such systems can take place in a few seconds when diffusion is accompanied by advection in a moving fluid medium.

There is a connection between the Lagrangian representation based on advected particles and the Eulerian representation using concentration fields. As in the case of pure advection the solution of the advection-diffusion equation can be given in terms of trajectories of fluid elements. Equation (2.6) can be generalized for the diffusive case using the Feynman-Kac formula (see e.g. Durrett (1996)) as

$$C(\mathbf{x}, t) = \left\langle C[\mathbf{r}(0), 0] + \int_0^t S[\mathbf{r}(t')]dt' \right\rangle_{\boldsymbol{\eta}} \qquad (2.33)$$

where the averaging is over an ensemble of trajectories satisfying the stochastic advection equation

$$\dot{\mathbf{r}} = \mathbf{v}(\mathbf{r}, t) + \boldsymbol{\eta}(t) \qquad (2.34)$$

subject to the terminal condition $\mathbf{r}(t) = \mathbf{x}$. $\boldsymbol{\eta}$ is a Gaussian white noise term with zero mean and delta-correlated in time

$$\langle \boldsymbol{\eta}(t)\boldsymbol{\eta}(t') \rangle = 2D\mathbf{I}\delta(t - t'). \qquad (2.35)$$

In the following we consider transport processes in a range of different types of flows of increasing complexity. In each case we first consider the purely advective transport that generates the trajectories of non-diffusive particles and then the combined effects of advection and diffusion on a continuous concentration field.

2.2   Steady two-dimensional flows

2.2.1   Advection along streamlines

As already mentioned the velocity field of incompressible two-dimensional flows can be represented by a scalar streamfunction $\psi(x, y)$ as

$\mathbf{v} = (\partial_y \psi, -\partial_x \psi)$. The motion of a fluid element is then described by an autonomous one-degree-of-freedom Hamiltonian dynamical system

$$\dot{x} = \frac{\partial \psi}{\partial y}, \quad \dot{y} = -\frac{\partial \psi}{\partial x} , \tag{2.36}$$

where the streamfunction plays the role of the Hamiltonian.

The rate of change of the time-independent streamfunction along the path of a fluid element, given by its Lagrangian derivative, vanishes

$$\frac{D\psi}{Dt} = \frac{\partial \psi}{\partial t} + \mathbf{v} \cdot \nabla \psi = \frac{\partial \psi}{\partial y}\frac{\partial \psi}{\partial x} - \frac{\partial \psi}{\partial x}\frac{\partial \psi}{\partial y} = 0 , \tag{2.37}$$

and then fluid particle trajectories are restricted to the streamlines defined by the level-curves $\psi(x,y) = \text{constant}$. This excludes the possibility of complicated trajectories in steady two-dimensional flows. The velocity field is everywhere tangent to the streamlines, thus there is no advective transport across the streamlines that act as transport barriers. The streamlines can be either closed curves on which the motion of fluid elements is periodic in time or open trajectories that go off to infinity in the inflow and outflow regions. An example of the typical structure of flow streamlines is shown in Fig. 2.2.

An important characteristic of advective transport is relative dispersion that describes how the separation of nearby fluid elements changes in time. The separation $\delta(t)$ between two fluid elements that are initially close to each other, $\delta(0) \ll L$, can be decomposed into tangential and normal components relative to the streamlines. For particles moving on closed streamlines the normal component oscillates periodically in time but the dominant asymptotic behavior comes from the tangential component δ_t that increases by the same amount after each cycle due to the small difference in the periods, $T(\psi)$, of the motion on nearby closed periodic orbits. Therefore

$$\delta_t((n+1)T) \approx \delta_t(nT) + |\mathbf{v}|\frac{dT}{d\psi}\nabla \psi \cdot \boldsymbol{\delta}_0 \tag{2.38}$$

where the last term is proportional to the normal component of the initial separation. Equation (2.38) remains valid while the separation is much smaller than the size of the orbit. Taking the limit

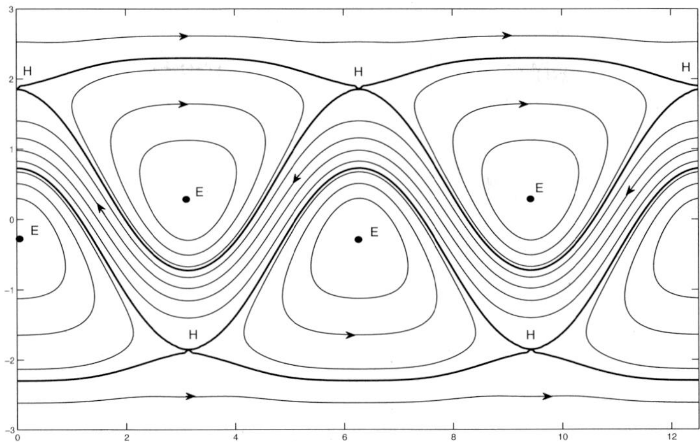

Figure 2.2: Streamlines of a meandering jet surrounded by recirculation zones showing typical structures in steady two-dimensional flows: elliptic (E) and hyperbolic (H) stagnation points, flow regions with open and closed streamlines and separatrices that connect hyperbolic points. The flow is defined by the streamfunction $\psi(x,y) = Cy - \tanh[(y - A\cos x)/(L\sqrt{1 + A^2 \sin x^2})]$.

of infinitesimally small initial separation the long time asymptotic behavior is thus linear in time

$$\lim_{t\to\infty}\lim_{\delta_0\to 0}\left(\frac{\delta(t)}{\delta_0}\right) \sim t. \tag{2.39}$$

This linear growth also holds for the length of material lines, e.g. representing a boundary between two fluid regions with different properties. As we will see later, time-dependent flows allow for much faster, accelerating growth of the length of material lines.

Typically, the flow field also contains stagnation points where $\mathbf{v}(\mathbf{x}^*) = 0$. The stagnation points can be classified according to the geometrical structure of the trajectories in their neighborhood. The trajectory of a fluid element around a stagnation point can be written as $\mathbf{r}(t) = \mathbf{x}^* + \boldsymbol{\delta}(t)$ and approximating the velocity field by a Taylor expansion around $\mathbf{x}^*$ the equation of motion for $\boldsymbol{\delta}(t)$ is

$$\dot{\boldsymbol{\delta}} = \nabla \mathbf{v}(\mathbf{x}^*) \cdot \boldsymbol{\delta}. \tag{2.40}$$

Taking into account the incompressibility of the flow, the eigenvalues of the velocity gradient tensor $\nabla \mathbf{v}$ can be written as

$$\lambda_{1,2} = \pm \left[\left(\frac{\partial v_x}{\partial x} \right)^2 - \left(\frac{\partial v_y}{\partial x} \right) \left(\frac{\partial v_x}{\partial y} \right) \right]^{1/2}. \tag{2.41}$$

Thus there are two qualitatively different cases. When the eigenvalues are complex, $\lambda_{1,2} = \pm i\omega$, $\mathbf{x}^*$ is an elliptic fixed point corresponding to a local maximum or minimum of the streamfunction. Such elliptic points are surrounded by closed circular orbits that stay in the close vicinity of $\mathbf{x}^*$. Therefore the area around the elliptic points remains isolated from the rest of the flow, thus preventing efficient mixing.

In the case of real eigenvalues, $\lambda_{1,2} = \pm \lambda$, $\mathbf{x}^*$ is called a *hyperbolic* fixed point, corresponding to a saddle point of the streamfunction, and it is of unstable character. It is located at the intersection of two special streamlines, called separatrices, which are the stable and unstable manifolds of the hyperbolic fixed point. These lines are defined as the set of points that approach the fixed point in the limits $t \to \pm\infty$, respectively. The motion around a hyperbolic point can be obtained as a solution of (2.40)

$$\boldsymbol{\delta}(t) = (\mathbf{n}_1^\dagger \cdot \boldsymbol{\delta}_0)e^{\lambda t}\mathbf{n}_1 + (\mathbf{n}_2^\dagger \cdot \boldsymbol{\delta}_0)e^{-\lambda t}\mathbf{n}_2 , \tag{2.42}$$

where $\mathbf{n}_{1,2}$ are the eigenvectors of the velocity gradient tensor at $\mathbf{x}^*$ and $\mathbf{n}_{1,2}^\dagger$ are the corresponding dual vectors that satisfy: $\mathbf{n}_1 \cdot \mathbf{n}_1^\dagger = \mathbf{n}_2 \cdot \mathbf{n}_2^\dagger = 1$ and $\mathbf{n}_1 \cdot \mathbf{n}_2^\dagger = \mathbf{n}_2 \cdot \mathbf{n}_1^\dagger = 0$. For long times the term with a positive exponent dominates (except for special points lying on the stable direction for which the coefficient of the growing exponential term is zero). Thus, almost all fluid particles leave the neighborhood of the fixed point along the unstable direction and the distance from $\mathbf{x}^*$ grows exponentially until the higher order terms in the Taylor expansion of the velocity field become important and the linear approximation breaks down. In the limit of infinitesimal initial separation the asymptotic behavior of the separation is:

$$\lim_{t\to\infty}\lim_{\delta_0\to 0}\left(\frac{\delta(t)}{\delta_0}\right)\sim e^{\lambda t}. \qquad (2.43)$$

Thus in the neighborhood of hyperbolic points the distance between fluid particles, or the length of material lines, grows exponentially in time that would lead to efficient mixing. However, since the fluid elements are quickly ejected from the vicinity of these isolated points they only have a short term transient effect with little influence on the global mixing properties of the flow.

2.2.2 Dispersion of diffusive tracers in steady flows

When considering large spatial and temporal scales the transport of a concentration field by advection and molecular diffusion can be be approximately described by a diffusion equation with an effective diffusion coefficient. The main question then is to find an expression for the effective diffusivity as a function of the flow parameters and molecular diffusivity. A range of this type of problems are discussed in the review by Majda and Kramer (1999). Here we consider two simple examples of this problem in the case of steady two-dimensional flows with open and closed streamlines, respectively.

Dispersion in a unidirectional shear flow

Advection and diffusion in a uni-directional shear flow of the form $\mathbf{v} = [u(y), 0]$ is a classical problem studied by Taylor (1953), considered as a simple model for the dispersion of a contaminant in a river or in a flow through a pipe (Fig. 2.3).

The concentration field satisfies the advection-diffusion equation

$$\frac{\partial C}{\partial t} + u(y)\frac{\partial C}{\partial x} = D\left(\frac{\partial^2 C}{\partial x^2} + \frac{\partial^2 C}{\partial y^2}\right). \qquad (2.44)$$

In the following we use an overbar to denote the mean of a quantity averaged along the transversal y direction, $\overline{A}(x) \equiv L^{-1}\int_L dy\, A(x, y)$, where L is the width of the channel. The velocity field can be

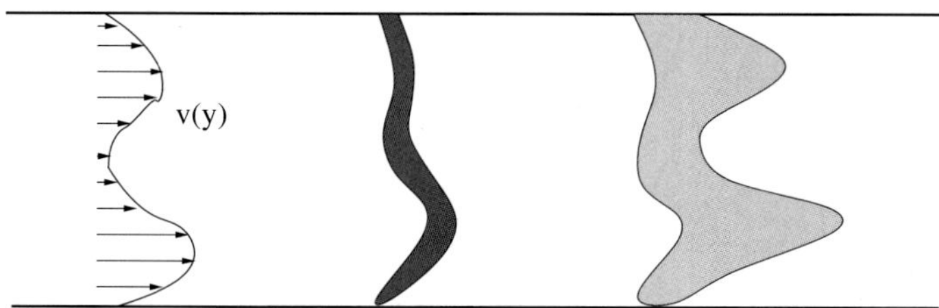

Figure 2.3: Longitudinal dispersion of a band of contaminant in a time-independent unidirectional shear-flow.

decomposed into a mean flow $\overline{u}$ and its deviation from the mean $u'(y) = u(y) - \overline{u}$. Averaging Eq. (2.44) along the y axis yields

$$\frac{\partial \overline{C}}{\partial t} + \overline{u}\frac{\partial \overline{C}}{\partial x} + \overline{u'\frac{\partial C'}{\partial x}} = D\frac{\partial^2 \overline{C}}{\partial x^2}, \tag{2.45}$$

where $C'(x, y) = C(x, y) - \overline{C}(x)$. Subtracting (2.45) from (2.44) gives

$$\frac{\partial C'}{\partial t} + u'\frac{\partial \overline{C}}{\partial x} + \overline{u}\frac{\partial C'}{\partial x} - \overline{u'\frac{\partial C'}{\partial x}} = D\left(\frac{\partial^2 C'}{\partial x^2} + \frac{\partial^2 C'}{\partial y^2}\right). \tag{2.46}$$

After sufficiently long time $t \gg L^2/D$, diffusion smoothes out the fluctuations in the y direction so that the dominant non-uniformity is along the longitudinal direction, i.e. $C' \ll \overline{C}$. Assuming that in the longitudinal direction advection is much faster than molecular diffusion the dominant balance in (2.46) is

$$u'\frac{\partial \overline{C}}{\partial x} = D\frac{\partial^2 C'}{\partial y^2}. \tag{2.47}$$

Thus, the shear flow acting on the y averaged concentration field continuously generates transversal fluctuations that are smoothed out by molecular diffusion. Integrating twice we obtain an expression for the fluctuating part of the concentration field

$$C' = \frac{1}{D}\frac{\partial \overline{C}}{\partial x} \int_0^y \int_0^{y'} u'(y'')dy''dy', \tag{2.48}$$

that can be used to calculate the term $\overline{u'\partial_x C'}$. Substituting it into (2.45) we obtain an equation for the y averaged concentration field

$$\frac{\partial \overline{C}}{\partial t} + \overline{u}\frac{\partial \overline{C}}{\partial x} = D_{eff}\frac{\partial^2 \overline{C}}{\partial x^2} \tag{2.49}$$

that has the form of a one-dimensional advection-diffusion equation where the drift along the channel is due to the mean flow and the dispersion along the channel is characterized by an effective diffusivity given by:

$$D_{eff} = D - \frac{1}{DL}\int_0^L u'(y)\left(\int_0^y \int_0^{y'} u'(y'')dy''dy'\right)dy. \tag{2.50}$$

Integrating by parts, the second term can be rewritten as

$$D_{eff} = D\left[1 + \frac{1}{D^2 L}\int_0^L \left(\int_0^y u'(y')dy'\right)^2 dy\right]. \tag{2.51}$$

Thus, we see that the shear flow enhances the dispersion along the channel, i.e. $D_{eff} > D$.

When the velocity profile is known, the effective diffusivity can be calculated explicitly. For example, in a plane Poiseuille flow with a parabolic velocity profile $u(y) = 4Uy(L-y)/L^2$ the diffusivity is

$$D_{eff} = D\left(1 + \frac{2U^2 L^2}{945 D^2}\right). \tag{2.52}$$

A similar expression can be obtained for an axially symmetric flow in a tube with a parabolic profile $v(r) = U(1 - r^2/R^2)$ as

$$D_{eff} = D\left(1 + \frac{U^2 R^2}{192 D^2}\right). \tag{2.53}$$

It is interesting to consider the dependence of D_{eff} on the molecular diffusion coefficient. For small molecular diffusivity, $D \ll UL$, i.e. large Péclet number, the second term dominates in (2.51) and $D_{eff} \sim U^2 L^2/D$ ($\sim DPe^2$). Thus weak molecular diffusion leads to large effective diffusivity. This somewhat counterintuitive result can be explained as follows. Longitudinal dispersion arises due to the different flow velocities across the channel. When the molecular diffusion across the streamlines is strong, the diffusively moving particles interchanged by the fluid elements sample the different flow velocities within a relatively short time, so that they are advected along the channel with almost the same mean velocity close to $\bar{u}$ and therefore the longitudinal dispersion is weak. However, when molecular diffusion is slow, the shear has a long time to stretch the concentration field creating large gradients before it is smeared out across the channel, allowing for stronger longitudinal dispersion by the flow.

This interpretation of the effective diffusion in terms of individual trajectories of an ensemble of particles advected by the flow and a superimposed random Brownian motion, as described by the stochastic advection equation (2.34), can be extended further. The characteristic time for molecular diffusion across the channel $\tau_D \sim L^2/D$ gives the correlation time of the longitudinal velocity experienced by a particle. Thus the longitudinal motion can be described as a collection of independent longitudinal displacements of typical length $U\tau_D$ over time intervals τ_D. Thus, for long times, $t \gg \tau_D$, the effective diffusion coefficient of such random walk can be estimated as $D_{eff} \sim (U\tau_D)^2/\tau_D \sim U^2 L^2/D$ that is consistent with (2.51) when $\mathrm{Pe} \gg 1$.

Dispersion in steady cellular flows

As a second example we consider dispersion in a cellular flow with no open streamlines. The flow is specified by the streamfunction

$$\psi(x,y) = UL \sin\left(\frac{2\pi x}{L}\right)\sin\left(\frac{2\pi y}{L}\right) \tag{2.54}$$

where U is the characteristic velocity and L is the size of the cells. This type of flow, in the geometry shown in Fig. 2.4, arises for example in Rayleigh-Bernard convection in a range of parameters corresponding to steady convection rolls.

Fluid elements are advected along closed circular streamlines, but the separatrices connecting the hyperbolic points inhibit long range advective transport from one cell to another (see Fig. 2.4). Thus

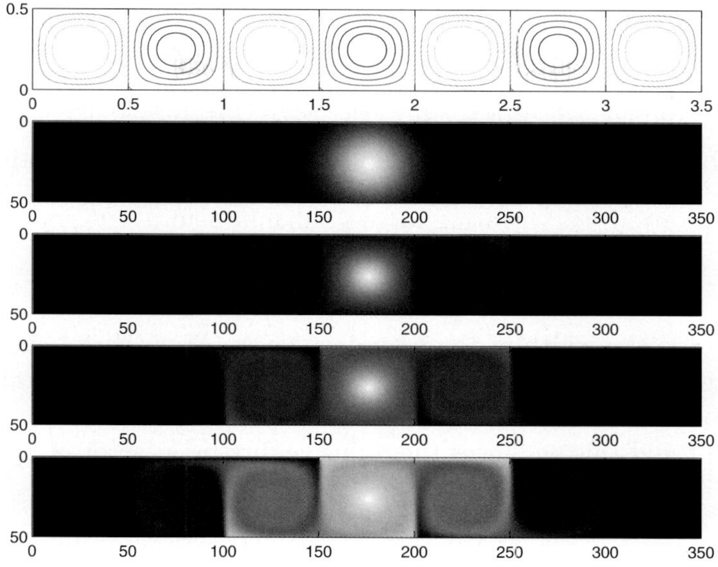

Figure 2.4: Streamlines of the steady cellular flow composed of an array of counter-rotating vortices (top row) and the spreading in time of a weakly diffusing passive concentration field. Time increases from top to bottom.

there is fast advective spreading of the concentration field within a cell, but transport between the cells is controlled by molecular diffusion only.

Using asymptotic expansion methods Shraiman (1987) and Rosenbluth et al. (1987) have shown that for large Pe the effective diffusivity is of the form

$$D_{eff} \sim (ULD)^{1/2} \sim D\mathrm{Pe}^{1/2}. \qquad (2.55)$$

Thus, the effective diffusivity can be much larger than the molecular one, as in the unidirectional shear flow case, although the dependence in D is different here. The origin of this scaling can be understood as follows. When the molecular diffusivity is small, i.e. $\mathrm{Pe} = UL/D \gg 1$, the characteristic time for advection within a roll

(L/U) is much smaller than the diffusion time over a cell, that is L^2/D. As a result of the dominant advective transport the concentration across a single cell is almost uniform and the concentration difference between two adjacent cells is restricted to a narrow boundary layer in the vicinity of the separatrix. In this region the dominant balance in the advection-diffusion equation is between advection along the separatrix and diffusion across the separatrix through the boundary layer

$$U\frac{\delta C}{L} \approx D\frac{\delta C}{l^2} \tag{2.56}$$

where l is the width of the boundary layer. From here l can be estimated as $l \sim (DL/U)^{1/2} = \mathrm{Pe}^{-1/2}L$. Thus advection concentrates the concentration gradient from the scale of the cell to the narrow region l and this enhances the diffusive flux by a factor of L/l

$$D\nabla C \sim D\frac{\delta C}{l} \sim D_{eff}\frac{\delta C}{L}, \tag{2.57}$$

which recovers the result $D_{eff} = (L/l)D = \mathrm{Pe}^{1/2}D$, valid for times much longer than the diffusion time over one cell, i.e. $t \gg L^2/D$, and for lengthscales much larger than the cell size L. This result has been verified experimentally by Solomon and Gollub (1988) who found good agreement with the theoretical prediction.

The two examples presented in this section illustrate that advection can significantly enhance transport in comparison to pure molecular diffusion, i.e. $D_{eff} \gg D$. Note, however, that the effective diffusion coefficient only characterizes the asymptotic long time behavior of the concentration field. On shorter times the fluid trajectories restricted to the streamlines do not allow for efficient mixing of the different fluid regions in these type of flows.

2.3 Advection in weakly time-dependent two-dimensional flows

In steady two-dimensional flows the spatial structure of the streamlines imposes strong restrictions on the advection of fluid elements.

However, this does not apply to the case when the flow is time-dependent. Here we consider the changes in advective transport arising from the time-dependence introduced as a small perturbation so that the streamfunction is of the form

$$\psi(x, y, t) = \psi_0(x, y) + \epsilon \psi_1(x, y, t) \tag{2.58}$$

where $\epsilon \ll 1$. In the simplest case the flow is periodic in time, i.e. $\psi_1(x, y, t) = \psi_1(x, y, t + nT)$. This is a typical situation for flows close to the steady-unsteady transition, for example at the onset of the oscillation of convection rolls in Rayleigh-Bernard convection or in the wake of an obstacle just above the critical Reynolds number at which the Kármán vortex street develops.

The advection problem is thus described by a periodically driven non-autonomous Hamiltonian dynamical system. In such case, besides the two spatial dimensions an additional variable is needed to complete the phase space description, which is conveniently taken to be the cyclic temporal coordinate, $\tau = t \mod T$, representing the phase of the periodic time-dependence of the flow. In time-dependent flows ψ is not conserved along the trajectories, hence trajectories are no longer restricted to the streamlines. The structure of the trajectories in the phase space can be visualized on a Poincaré section that contains the intersection points of the trajectories with a plane corresponding to a specified fixed phase of the flow, τ_0. On this stroboscopic section the advection dynamics can be defined by the stroboscopic Lagrangian map

$$\mathbf{r}[(n + 1)T + \tau_0] = \boldsymbol{\Phi}_T[\mathbf{r}(nT + \tau)] \tag{2.59}$$

that gives the new positions of fluid elements after one full period of the flow.

The dynamics of such systems is described by the Kolmogorov-Arnold-Moser theory of nearly integrable conservative dynamical systems (see e.g. Ott (1993)). For $\epsilon = 0$ the fluid elements move along the streamlines and the trajectories in the phase space form tubes parallel to the time axis. Due to the periodicity in the temporal direction these tubes form tori that fill the whole phase space and are invariant surfaces for the motion of the fluid elements. Each torus can be characterized by a rotation number, that is the ratio of the period of the motion around the closed streamline and the period of the perturbation $R = T_\Psi/T$. Almost all tori have an irrational rotation number and contain quasiperiodic orbits that fill their surface and never exactly return to the same point in the phase space. There is also an infinite set of tori with rational rotation numbers $R = p/q$, with p and q integers, that contain an infinite number of closed periodic orbits with period qT. They are the *resonant tori* with respect to a given periodic perturbation.

The main question is what happens with these invariant surfaces when the streamfunction has a small time-periodic component, $0 < \epsilon \ll 1$. Are there any invariant surfaces preserved when the perturbation is small or they disappear for arbitrarily small perturbations and the orbits may wander anywhere in the phase space? The answers are given by important theorems from the field of Hamiltonian dynamical systems.

According to the Poincaré-Birkhoff fixed point theorem all resonant tori break up for arbitrarily small perturbations. If the rotation number is p/q the perturbation leaves q pairs of hyperbolic and elliptic periodic orbits. The unstable hyperbolic orbits are embedded in a layer filled by aperiodic, chaotic orbits that do not stay on an invariant surface, but cover a finite non-zero volume of the phase space in a chaotic layer around the original resonant torus. The elliptic points, however, are wrapped around by new concentric tori that form islands of regular orbits within the chaotic band (Fig. 2.5).

Although in a weakly time-dependent flow all resonant tori disappear together with some of the nearly resonant tori around them, the Kolmogorov-Arnold-Moser theorem ensures that infinitely many invariant surfaces survive a small perturbation. For sufficiently small ϵ the remaining invariant surfaces formed by quasiperiodic orbits, so called KAM tori, still occupy a non-zero volume of the phase space. The condition for a torus to survive a given perturbation is that its rotation number should be sufficiently far from any rational number so that the inequality

$$\left| R - \frac{p}{q} \right| > \frac{K(\epsilon)}{q^{2.5}} \tag{2.60}$$

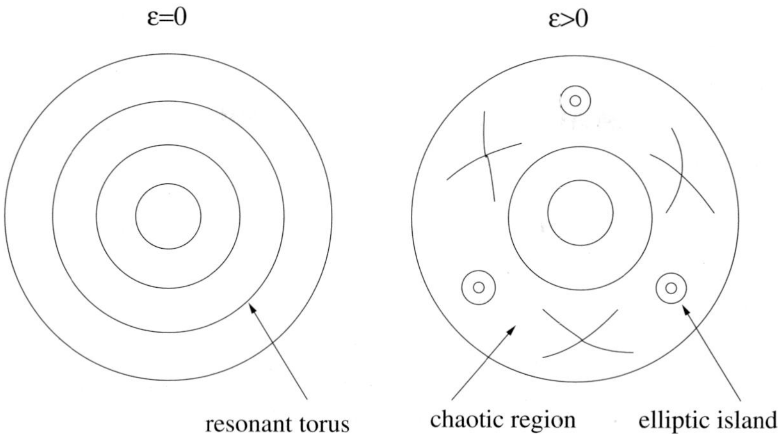

Figure 2.5: Sketch of the typical structures associated with the break-up of a resonant torus on the Poincaré section in a weakly time-periodic flow producing a set of new elliptic and hyperbolic points.

holds for any pair of integers p, q. $K(\epsilon)$ is a number which depends on the intensity of the perturbation ϵ. Thus some far-from-resonant tori may persist even when the time-dependence is relatively strong, while around the most resonant tori, i.e. with p, q small integers, a prominent chaotic region develops already at very small perturbations.

Thus some of the fluid elements move on aperiodic chaotic trajectories and others on quasiperiodic orbits. The quasiperiodic orbits are invariant surfaces in the phase space that form the boundaries of the chaotic layers and limit the motion of the chaotic trajectories. There is a similar structure around each elliptic periodic orbit resulting from broken resonant tori that are also surrounded by invariant tori forming isolated islands inside the chaotic region.

Thus, the time-dependence of the flow generates chaotic trajectories that will enhance the mixing of fluid within these regions. However, the KAM tori formed by the remaining quasiperiodic orbits separate the domain into a set of disconnected regions with no advective transport between them. Therefore, when the time-dependence is weak the fluid is only mixed within narrow layers around the resonant streamlines of the original time-independent flow. The areas

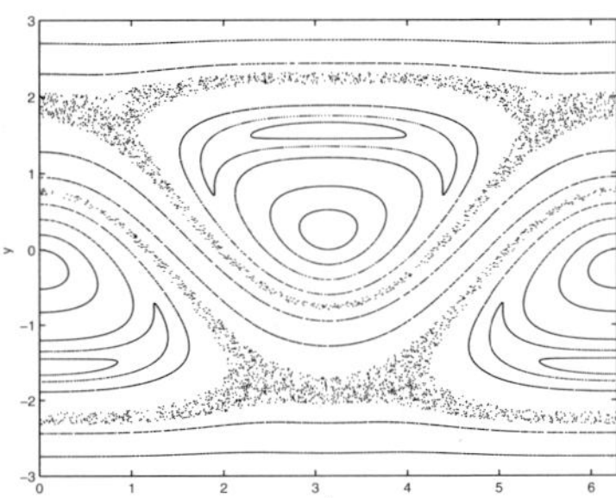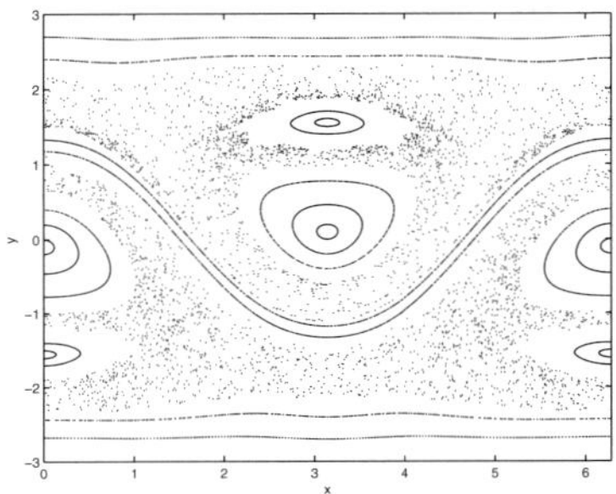

Figure 2.6: Poincaré sections of particle trajectories in a time-periodic jet flow. The streamfunction is the same as in Fig. 2.2 with $A = A_0 + \epsilon \sin \omega t$ where $\epsilon = 0.02$ (left) and $\epsilon = 0.1$ (right). Dots indicate chaotic regions separated by trajectories on KAM tori, which appear as continuous lines.

that are most sensitive to time-periodic perturbations are around the separatrices where broad chaotic regions appear already for small perturbations (see Fig. 2.6). As the perturbation is increased, more and more invariant surfaces disappear and the irregular chaotic orbits can access larger flow regions resulting in more efficient mixing. Eventually only very small tori remain that have no significant effect on transport. In such globally chaotic systems the fluid elements can approach almost every point within the domain.

A mathematically more precise concept associated to this property is *ergodicity*. The stroboscopic map Φ_T is ergodic if all of its invariant sets (i.e. those that satisfy $\Phi_T(A) = A$) have zero measure or their complement has zero measure. A stronger property is *mixing* that requires that for any two sets A and B the measure μ (area or volume) of the intersection satisfies $\mu(\Phi_T^n(A) \cap B) \to \mu(A)\mu(B)$ for large n. This means that after many applications of the map Φ_T the set A will be uniformly spread over the whole domain so its overlap with another set B is proportional to the area of B (for example as shown by the black and white regions in Fig. 2.9 on page 49). Note that every mixing system is ergodic, but the opposite is not true. A quasiperiodic motion on a torus is ergodic, but it is not mixing in the above sense.

2.4   Chaotic advection in three dimensions

In three-dimensional flows the velocity field cannot be defined through a streamfunction, therefore the advection of fluid elements does not have the simple Hamiltonian structure as in two dimensions. One significant result on mixing in three dimensions is related to the existence of invariant surfaces in steady inviscid flows (Arnold, 1965). The velocity field of such flows is a solution of the time-independent Euler equation

$$\mathbf{v} \cdot \nabla \mathbf{v} = -\nabla p, \tag{2.61}$$

that can be rewritten as

$$\mathbf{v} \times (\nabla \times \mathbf{v}) = \nabla \alpha, \tag{2.62}$$

where $\alpha = p + |\mathbf{v}|^2/2$ is the Bernoulli function. From (2.62) follows that the gradient of α is perpendicular to the velocity vector, which is therefore tangent to the iso-surfaces defined by $\alpha = constant$. Thus the motion of fluid trajectories is restricted to these invariant surfaces. According to the Poincaré-Bendixon theorem (Jackson, 1989, Vol. 1), a time-independent flow on a two-dimensional invariant surface cannot be chaotic, therefore all bounded trajectories must be closed periodic orbits. The only exception to this is the case when the vorticity and velocity vectors are parallel to each other. In such so called *Beltrami flows* that satisfy $\boldsymbol{\omega} = a\mathbf{v}$, α is constant everywhere in space and does not impose any restriction on the trajectories. Therefore, Arnold suggested that advection in Beltrami flows may produce three-dimensional chaotic trajectories.

An example of a Beltrami flow is the ABC flow, named after Arnold, Beltrami and Childress. It is defined by the velocity field

$$v_x = A \sin z + C \cos y \tag{2.63}$$
$$v_y = B \sin x + A \cos z \tag{2.64}$$
$$v_z = C \sin y + B \cos x \tag{2.65}$$

which is a solution of the time-independent Euler equation. Dombre et al. (1986) have shown that when all three parameters A, B and C are non-zero the fluid elements move chaotically in regions separated by invariant surfaces that are similar to the KAM tori of unsteady two-dimensional flows.

Beltrami flows are rather special solutions of the Euler equation, which only applies to inviscid fluids. Time-independent three-dimensional viscous flows, in general, could generate chaotic motion of the fluid elements. Special cases with invariant surfaces inhibiting chaotic motion are flows with axial or rotational symmetry, e.g. a steady flow in a pipe or flow generated by a symmetrically rotating impeller. When a small perturbation breaks these symmetries the transition to chaotic advection can be observed via the same scenario as in the case of weakly time-periodic flows. Note that in this case the phase space of the advection dynamics coincides with the three-dimensional physical space, so structures like KAM tori, chaotic bands and elliptic islands can be made directly visible in experiments. This was demonstrated by Fountain et al. (1998) in a flow produced by a rotating disc impeller placed in a cylindrical tank. The flow structures were visualized by injecting fluorescent dye at certain points and illuminating the tank with a laser sheet that creates a Poincaré section of the advection dynamics. When the axis of the cylinder and of the impeller coincide, all fluid elements move quasiperiodically on toroidal invariant surfaces. Changing the angle between the two axis destroys the rotational symmetry and the formation of thin chaotic layers and elliptic islands can be observed, as predicted by the Kolmogorov-Arnold-Moser theory (Fig. 2.7).

The effect of small time-periodic perturbations in three-dimensional flows with regular non-chaotic dynamics has been studied by Cartwright et al. (1995, 1996). They classified non-chaotic steady three-dimensional flows in terms of action and angle type variables (Feingold et al., 1988) as: (*i*) *action-angle-angle* flows in which the motion takes place on tori defined by a constant action I and the particles position is characterized by two angle variables: $\dot{I} = 0, \dot{\phi} = \omega_1(I), \dot{\theta} = \omega_2(I)$; and (*ii*) *action-action-angle* flows in which fluid particles move on a two-parameter family of invariant curves: $\dot{I} = 0, \dot{J} = 0, \dot{\phi} = \omega(I, J)$. The effect of weak time-periodic perturbations on regular action-angle-angle type flows is very similar to the case of two-dimensional flows discussed in the previous section. When the perturbation is sufficiently small most invariant tori survive. The

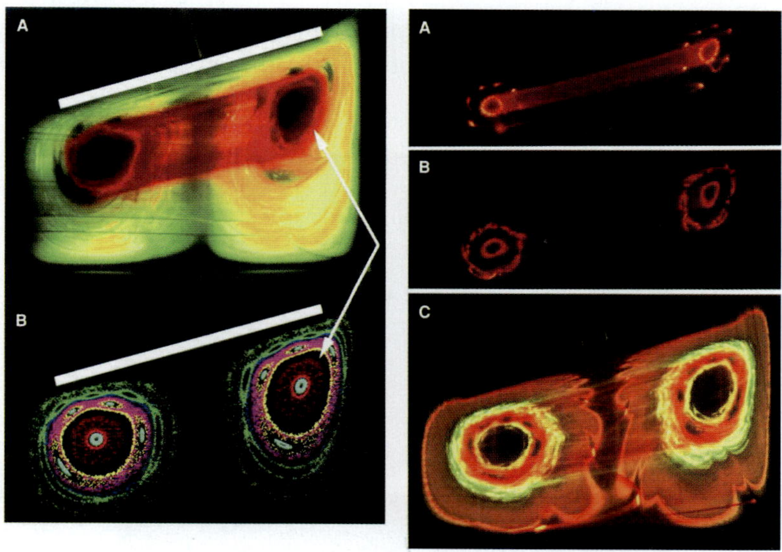

Figure 2.7: KAM tori and elliptic islands visualized by fluorescent dye in an experiment with a steady three-dimensional flow in a viscous fluid (from Fountain et al. (1998)).

resonant tori defined by $p\omega_1 = q\omega_2$ for some integers p, q, are destroyed by any small perturbation and they are replaced by a set of stable and unstable invariant orbits with chaotic motion around the unstable ones and new invariant tori surrounding the stable ones. The surviving tori act as transport barriers that confine the motion of the chaotic trajectories to small volumes when the perturbation is small.

However, the effect of a small perturbation in action-action-angle type flows is quite different. The two-parameter family of invariant cycles coalesce into invariant tori that are connected by resonant sheets defined by the $\omega(I_1, I_2) = 0$ condition. The consequence of this is that contrary to action-angle-angle flows in this case a trajectory can cover the whole phase space and no transport barriers exist. Thus, in this type of flows global uniform mixing can be achieved for arbitrarily small perturbations. This type of *resonance induced dispersion* has been demonstrated numerically in a low-Reynolds number Couette flow between two rotating spheres by Cartwright et al.

(1995) and later was confirmed experimentally by Solomon and Mezic (2003) using a weakly time-periodic three-dimensional flow formed by a chain of oscillating vortices.

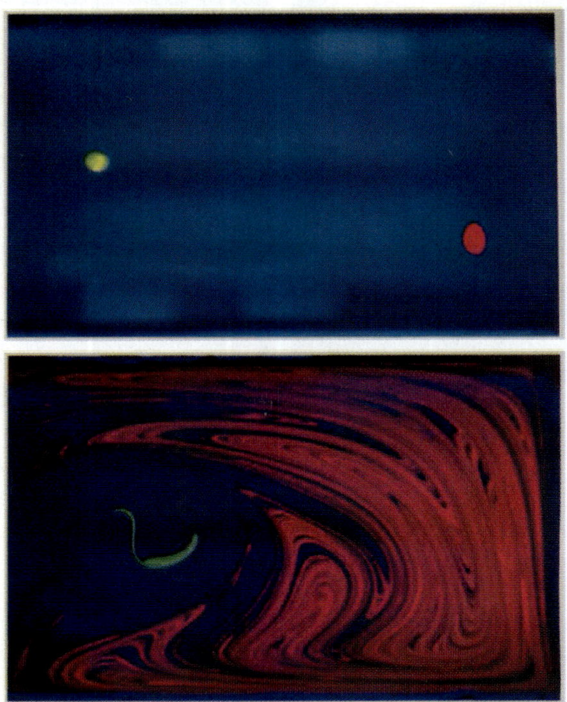

Figure 2.8: Advection of two blobs of dye in a periodically oscillating viscous flow placed in elliptic (green) and chaotic (red) flow regions. Experiment from Ottino (1989b).

2.5　Dispersion by chaotic advection

The relevance of chaotic motion of fluid elements for the mixing of fluids was first recognized by Aref (1984). The key property of the chaotic orbits is their sensitivity to small changes in the initial conditions. Since small perturbations grow fast in time fluid elements initially close to each other end up later in very different regions. This also holds for the time reversed chaotic advection dynamics, so

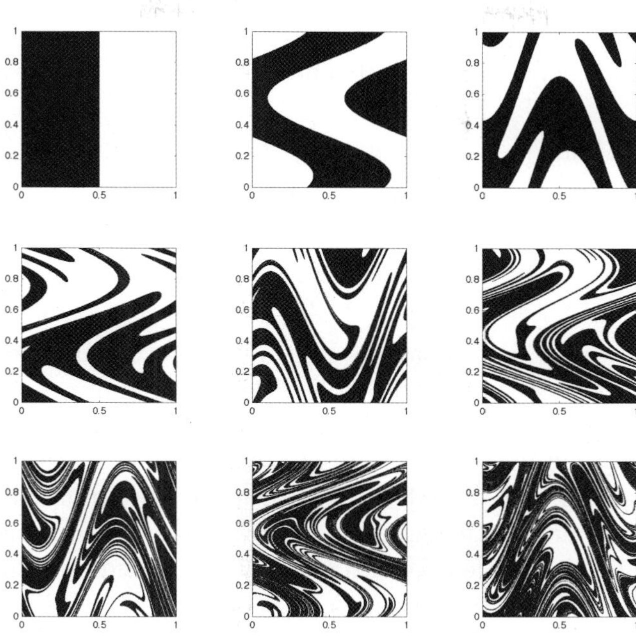

Figure 2.9: Blending of the left and right halves of the domain, marked black and white, by chaotic advection in the alternating sinusoidal shear flow (Eq. (2.66)), $\alpha \equiv UT/L = 0.8$. Time runs from left to right and then from top to bottom.

that nearby fluid elements in a chaotic region of the flow originate from very different regions. Clearly, such chaotic orbits are good for mixing as they efficiently bring together fluid parcels with different properties (Aref, 2002; Wiggins and Ottino, 2004). Mixing by chaotic advection was demonstrated experimentally by Chaiken et al. (1986) in a simple periodically forced laminar flow and further experimental work has been done by Ottino and co-workers (Ottino (1989a, 1990), Fig. 2.8). More recently various types of microfluidic devices have been designed that are able to generate chaotic advection and achieve good mixing even at microscopic scales (Stroock et al., 2002; Ottino and Wiggins, 2004; Squires, 2005).

A simple numerical example of chaotic advection in a piecewise steady sinusoidal shear flow periodically alternating along the x and y direction is shown in Fig. 2.9, where the velocity field is defined as

$$
\begin{aligned}
v_x &= U \sin\left(\frac{2\pi y}{L} + \phi\right), v_y = 0 \quad \text{for} \ \ nT < t < nT + \frac{T}{2} \\
v_x &= 0, v_y = U \sin\left(\frac{2\pi x}{L} + \phi\right) \quad \text{for} \ \ nT + \frac{T}{2} < t < (n+1)T.
\end{aligned}
$$

$$(2.66)$$

Due to the periodic spatial structure this represents a closed flow system with periodic boundary conditions on the unit square. Note that the trajectories of the fluid elements only depend on the non-dimensional parameter $\alpha = UT/L$. When α is small the flow is weakly chaotic and the advective transport is dominated by KAM tori that confine chaotic transport to narrow bands. For larger values, i.e. $\alpha \simeq \mathcal{O}(1)$, only a few small regular elliptic islands remain within a large connected chaotic region. Transport barriers and elliptic islands can be eliminated completely when the flow is aperiodic. This can be achieved by setting the phase ϕ to be a random variable that takes on different values in each flow period, thus generating a temporally aperiodic but spatially smooth laminar flow. Since quasiperiodic orbits are not possible in an aperiodic flow, in this case the whole domain is covered by a single fully connected mixing region in which chaotic advection dynamics is qualitatively similar to that in the deterministic time-periodic flow.

2.5.1   The Lyapunov exponent

The non-dimensional parameter $\alpha = UT/L$ of the sine-flow that controls the extent of chaotic advection can be interpreted as a ratio of two characteristic timescales. One of them is the typical advection time over the characteristic lengthscale of the velocity field L/U. This is a property of the instantaneous velocity field and would be the same for a steady flow. Therefore it can not characterize the dynamics of chaotic mixing. For a time-periodic velocity field another timescale is the period of the flow. In the case of an aperiodic time-dependent flow an analogous timescale can be defined as the correlation time of the velocity field. From the point of view of chaotic advection the important quantity that characterizes the chaotic mixing process is the ratio of these two timescales.

A more exact quantitative characterization of the chaotic advection can be given by considering the relative dispersion of fluid particles. Let us consider two fluid elements moving on trajectories $\mathbf{r}(t)$ and $\mathbf{r}'(t)$. When the distance between them is small compared to the characteristic lengthscale of the velocity field (L) the velocity difference can be approximated by Taylor expansion and the separation $\boldsymbol{\delta}(t) = \mathbf{r}'(t) - \mathbf{r}(t)$ satisfies the equation

$$\dot{\boldsymbol{\delta}} = \mathbf{v}[\mathbf{r}'(t), t] - \mathbf{v}[\mathbf{r}(t), t] \approx \nabla\mathbf{v}[\mathbf{r}(t), t]\boldsymbol{\delta}. \tag{2.67}$$

The solution of (2.67) can be written in the form

$$\boldsymbol{\delta}(t) = \mathbf{M}_t(\mathbf{x}_0)\boldsymbol{\delta}(0), \tag{2.68}$$

where the matrix $\mathbf{M}$ is the deformation gradient tensor whose elements are related to the Lagrangian map by $M_{ij} = \partial(\boldsymbol{\Phi}_t)_i/\partial r_j$, evaluated at the initial condition $\mathbf{r}(t = 0) = \mathbf{x}_0$. Thus the distance between the two particles is

$$|\boldsymbol{\delta}(t)|^2 = \boldsymbol{\delta}_0^T \mathbf{M}_t^T \mathbf{M}_t \boldsymbol{\delta}_0. \tag{2.69}$$

The Oseledec theorem (Eckmann and Ruelle, 1985) implies that under rather general conditions the eigenvectors of $\mathbf{M}_t^T(\mathbf{x}_0)\mathbf{M}_t(\mathbf{x}_0)$ converge, as $t \to \infty$, to a set of Lyapunov vectors $\{\mathbf{v}_i\}$. Since $\mathbf{M}_t^T\mathbf{M}_t$ is symmetric and positive definite the Lyapunov vectors form an orthonormal set and the eigenvalues Λ_i^t are all positive. If the initial separation is chosen to be oriented along one of the Lyapunov vectors, $\boldsymbol{\delta}_0 = \delta_0 \mathbf{v}_i$ with $\delta_0 \ll L$, the distance between the particles at a later time t is

$$|\mathbf{M}_t\delta_0\mathbf{v}_i| = (\delta_0 \mathbf{v}_i^T \mathbf{M}_t^T \mathbf{M}_t \delta_0 \mathbf{v}_i)^{1/2} = \delta_0(\Lambda_i^t)^{1/2}. \tag{2.70}$$

The linear approximation for the velocity differences in (2.67) is only valid while $|\boldsymbol{\delta}(t)| \ll L$ so this will only hold for arbitrary long time if we first take the limit $\delta_0 \to 0$. Then the set of Lyapunov exponents can be defined as the growth rate of the inter-particle distance as

$$\lambda_i = \lim_{t\to\infty} \lim_{\delta_0\to 0} \frac{1}{t} \ln\left(\frac{|\mathbf{M}_t\delta_0\mathbf{v}_i|}{\delta_0}\right) = \lim_{t\to\infty} \frac{\ln \Lambda_i^t}{2t}. \tag{2.71}$$

In general λ_i may depend on the initial condition $\mathbf{x}_0$. But for ergodic systems it has the same value for almost all initial positions (Eckmann and Ruelle, 1985), i.e. everywhere except in a set of measure zero. The characteristic signature of chaotic advection is that at least one of the Lyapunov exponents is positive, representing exponential growth of the distance separating the two particles.

The Lyapunov exponents can be either positive or negative indicating exponentially diverging or converging pairs of trajectories. The convention is to order them in a non-increasing sequence. Thus we have $\lambda_1 \geq \lambda_2 \geq \lambda_3 \geq \cdots \geq \lambda_d$, where d is the space dimensionality. Incompressibility implies that $\det(\mathbf{M}) = 1$ from which follows that the product of the eigenvalues Λ_i^t is unity and the sum of the Lyapunov exponents is zero. In three dimensions, if there are non-zero Lyapunov exponents, then necessarily $\lambda_1 > 0$ and $\lambda_3 < 0$ while λ_2 can be either positive or negative. In two-dimensional incompressible flows the two Lyapunov exponents must be equal in magnitude and of opposite sign, $\lambda_2 = -\lambda_1$, or they are both zero in the special case of non-hyperbolic systems. The constraint imposed by incompressibility on the Lyapunov exponents can be understood by considering a small spherical fluid region in the flow that by advection evolves into an elongated ellipsoid as it contains particle pairs that move away from each other at a rate given by the positive Lyapunov exponent. However, to conserve the volume of the fluid this must be compensated by convergence of some fluid elements in other directions that is characterized by the negative Lyapunov exponents.

When the direction of the initial separation does not coincide with one of the vectors $\mathbf{v}_i$, $\boldsymbol{\delta}_0$ can be projected on the orthonormal basis formed by the Lyapunov vectors (at point $\mathbf{x}_0$) as $\boldsymbol{\delta}_0 = \sum c_i \mathbf{v}_i$, that under advection by the flow becomes

$$\boldsymbol{\delta}(t) = \sum_i c_i \mathbf{M}_t(\mathbf{x}_0)\mathbf{v}_i. \tag{2.72}$$

The length of the vectors $\mathbf{M}_t \mathbf{v}_i$ grows or decreases exponentially in time as $\exp(\lambda_i t)$. After long time the sum (2.72) will be dominated by the term corresponding to the largest Lyapunov exponent. Thus, for almost all initial directions of the separation vector the distance between two fluid particles increases exponentially as

$$\lim_{t \to \infty} \lim_{\delta_0 \to 0} \frac{|\boldsymbol{\delta}(t)|}{|\boldsymbol{\delta}_0|} \sim e^{\lambda_1 t}, \tag{2.73}$$

and the direction of $\boldsymbol{\delta}(t)$ converges to the direction of $\mathbf{M}(\mathbf{x}_0)_t \mathbf{v}_1$ regardless of the initial orientation of $\boldsymbol{\delta}_0$. The only exception is when $\boldsymbol{\delta}_0$ is exactly perpendicular to $\mathbf{v}_1$ so that $c_1 = 0$. In this case the first term in (2.72) vanishes and the dominant contribution comes from the second term.

Thus the dominant Lyapunov vector identifies a well defined geometrical structure by assigning a unique asymptotic direction $\mathbf{M}(\mathbf{x}_0)_t \mathbf{v}_1$ to every point $\mathbf{x} = \boldsymbol{\Phi}_t(\mathbf{x}_0)$ in the flow. Fieldlines everywhere tangent to this direction form the *unstable foliation* of the chaotic advection dynamics, that indicates the locally dominant direction of the fastest separation of fluid particles or the direction of stretching of small fluid volumes as they are deformed by the flow. Material lines or contours advected by the flow can be considered as a collection of infinitesimal segments that independently of their initial orientation evolve into a curve that is everywhere aligned to the unstable foliation. As the vectors $\mathbf{M}_t \mathbf{v}_1$ evolve in time, the unstable foliation also changes following the time-dependence of the velocity field. Note that the unstable foliation is a property of the Lagrangian time-evolution and depends on the time-history of the flow, so that in general it is very different from the streamlines that represent the instantaneous directionality of the velocity field.

Similar direction fields can be associated to the subdominant Lyapunov exponents via the vectors $\mathbf{M}_t \mathbf{v}_i$. In the case of the most negative Lyapunov exponent this represents a *stable foliation* associated to the local direction of the fastest convergence of fluid particles. If diffusion is ignored, the concentration of advected quantities is conserved within fluid parcels, therefore the convergence of fluid elements in the direction of the stable foliation leads to concentration gradients increasing exponentially in time. Explicitly:

$$|\nabla C| \approx \frac{\delta C}{|\mathbf{M}_t \mathbf{v}_d|} \sim e^{-\lambda_d t} \tag{2.74}$$

where $\lambda_d < 0$, with $d = 2$ or $d = 3$ in two-dimensional or three-dimensional flows, and the direction of the stable foliation coincides with the direction of the gradient of the advected concentration field. The increase of the concentration gradients is finally limited by molecular diffusion. Based on the same arguments, in the direction of the unstable foliation the concentration field remains smooth. Thus concentrations develop a characteristic filamental structure with strong local anisotropy. Roughly speaking, dynamical systems for which this structure of stable and unstable directions is found everywhere, without tangencies, are called *hyperbolic.*

A numerical example of the evolution of a material line in the two-dimensional chaotic flow (2.66) is shown in Fig. 2.10. The initially circular contour first becomes an elongated ellipse with its axes expanding and contracting as $\sim \exp(\pm\lambda_1 t)$ while the enclosed area stays constant. The length in the expanding direction quickly becomes comparable to the characteristic scale of the velocity field and the linear approximation (2.67) breaks down in the expanding direction. Consequently, the initial ellipsoid becomes distorted into a long thin folded filament oriented along the unstable foliation whose width continues to shrink exponentially.

The exponential dispersion is only valid until the distance between the particles is smaller than the characteristic length scale of the flow. If initially there is a finite distance, δ_0, between the two particles, $0 < \delta_0 \ll L$, the exponential growth $\delta(t) \sim \delta_0 \exp(\lambda_1 t)$ saturates at around $t_s \simeq \ln(L/\delta_0)$. If the advection takes place within a bounded domain, then after this time the distance between the two particles fluctuates chaotically and is comparable to the size of the domain. In the case of an open unbounded system, e.g. in a spatially periodic velocity field, the dispersion becomes diffusive and $\delta(t) \sim t^{1/2}$ for large t.

Finite-time Lyapunov exponents

The set of asymptotic Lyapunov exponents does not fully capture all transport properties of the chaotic advection. A more complete description is given by the distribution of finite-time Lyapunov exponents. These quantities are defined as in (2.71) but for finite t. An alternative but related quantity, the finite-size Lyapunov exponent,

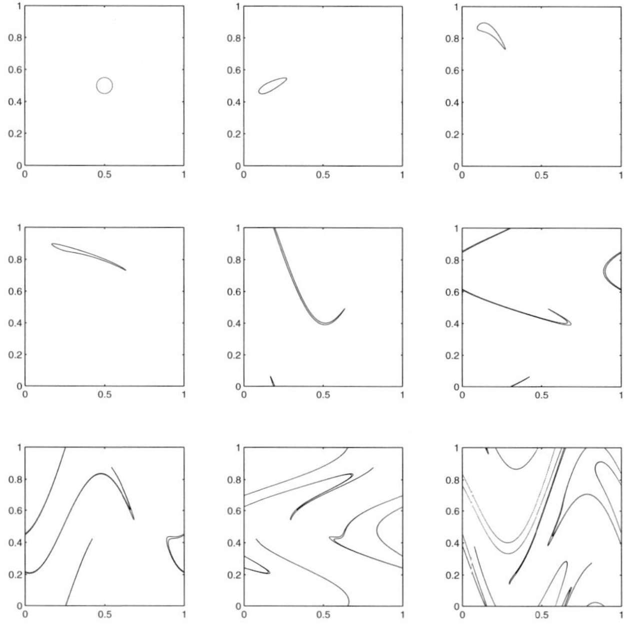

Figure 2.10: An initially circular contour advected by the sine-flow in Eq. (2.66). Time increases from left to right and then from top to bottom.

will be discussed in Sect. 2.8.1. The leading Lyapunov exponent is the asymptotic average stretching rate encountered by fluid elements along their trajectory in the limit $t \rightarrow \infty$, and it is independent of initial conditions in ergodic systems. In contrast, the finite-time averages of the stretching rate are non-uniform in space, so that estimates of the Lyapunov exponents based on the growth rate of an infinitesimal separation from finite-time trajectories yields a range of different values depending on the initial position, that may be different from the unique asymptotic Lyapunov exponent λ^{∞}. In chaotic systems, for times much larger than the correlation time of the chaotic dynamics, the distribution of finite-time Lyapunov exponents is approximately of the form (Ott, 1993; Bohr et al., 1998):

$$P(\lambda, t) \sim t^{1/2} e^{-G(\lambda)t} \qquad (2.75)$$

where the entropy function $G(\lambda) \geq 0$ is convex and is equal to zero at its minimum, which corresponds to the asymptotic Lyapunov exponent, i.e. $\min[G(\lambda)] = G(\lambda^\infty) = 0$. As t increases the finite-time Lyapunov exponents yield closer approximations to the asymptotic value and strong deviations become very rare. The probability density decays everywhere except around its maximum approaching a delta function $\delta(\lambda - \lambda^\infty)$ (see Fig. 2.11). In the region close to the asymptotic Lyapunov exponent, i.e. neglecting large deviations, $G(\lambda)$ can be approximated by a quadratic function

$$P(\lambda, t)t^{1/2} \sim e^{-\frac{(\lambda-\lambda^\infty)^2}{2\Delta}t}, \quad \text{where } \Delta = \frac{1}{G''(\lambda^\infty)} \qquad (2.76)$$

which is a Gaussian distribution with a variance Δ/t that decreases in time as the distribution becomes more and more peaked at λ^∞.

The proportion of fluid elements experiencing a particular anomalous value of the Lyapunov exponent $\lambda \neq \lambda^\infty$ decreases in time as $\exp(-G(\lambda)t)$. In the infinite-time limit, in agreement with the Oseledec theorem, they are limited to regions of zero measure that occupy zero volume (or area in two dimensions), but with a complicated geometrical structure of fractal character, to which one can associate a non-integer fractional dimension. Despite their rarity, we will see that the presence of these sets of untypical Lyapunov exponents may have consequences on measurable quantities. Thus we proceed to provide some characterization for their geometry.

The standard procedure to assign a dimension to a fractal set (see e.g. Ott (1993)) is to cover it with smaller and smaller objects, line segments, squares or cubes, depending on the dimensionality of the space, of linear size l. If in the limit $l \to 0$ the number of these objects needed to cover the set follows a power law

$$N(l) \sim l^{-D_f}, \qquad (2.77)$$

then D_f is the fractal dimension of the set. For a single point $N(l) = 1$, and for a line segment of length l_0, $N(l) \approx l_0/l$, thus in these simple

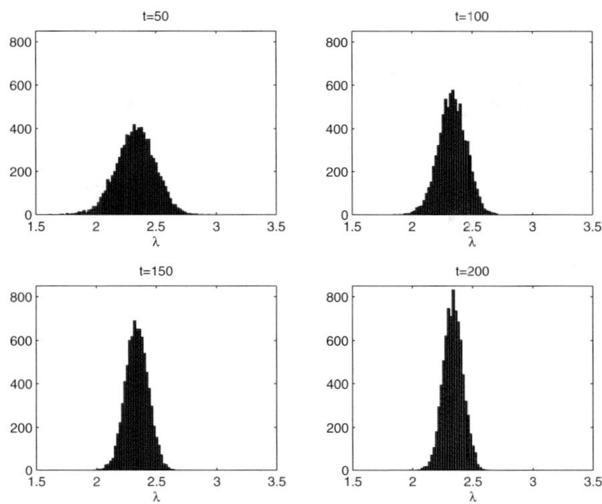

Figure 2.11: Histogram of the finite-time Lyapunov exponents calculated from 8000 trajectories in the chaotic sine-flow of Eq. (2.66) and Fig. 2.9, for $t = 50, 100, 150$ and 200. The distribution becomes increasingly concentrated around λ^{∞}.

cases D_f coincides with the conventional integer dimensions of 0 and 1, respectively.

For calculating the dimension of the set of spatial locations which at each time are occupied by trajectories with non-typical values of the asymptotic Lyapunov exponent we cover these locations with objects of size which can be estimated from the dynamics: Since the proportion of fluid elements experiencing a finite-time average stretching $\lambda \neq \lambda^{\infty}$ decays as $\exp(-G(\lambda)t)$, and in an incompressible flow the area of fluid element remains constant (we refer to the two-dimensional situation for simplicity), the total area covered by these fluid elements decreases also as $A_\lambda(t) \sim \exp(-G(\lambda)t)$. This area is stretched by the chaotic dynamics locally characterized by λ, so that its characteristic width shrinks as $w_\lambda(t) \sim \exp(-\lambda t)$. The number of boxes of size $l = w_\lambda$ needed to cover such set of fluid elements can be estimated as

$$N_\lambda(l) \simeq \frac{A_\lambda}{(w_\lambda)^2} \sim e^{[2\lambda - G(\lambda)]t} \sim l^{[G(\lambda)/\lambda] - 2}. \qquad (2.78)$$

Therefore, in the $t \to \infty$ limit, fluid elements experiencing a certain particular anomalous value of the Lyapunov exponent form a fractal set with dimension

$$D_f(\lambda) = 2 - \frac{G(\lambda)}{\lambda}. \tag{2.79}$$

The object composed of all these fractal sets is called a *multifractal*. The area covered by the anomalous sets is zero, but we will see that they can have a non-negligible contribution to certain flow characteristics defined by averages over the whole set of flow trajectories. As an example we consider the growth rate of the length of material lines under chaotic advection. The contour length can be expressed as a sum of infinitesimal line segments for which the linear approximation remains valid for arbitrary long time. Since the stretching at different points along the contour is typically non-uniform we have to take into account the distribution of finite time Lyapunov exponents for the finite-time trajectories. Thus the total length of the contour at a finite-time can be written as

$$l(t) \sim l_0 \int e^{\lambda t} P(\lambda, t) d\lambda \sim l_0 \int e^{(\lambda - G(\lambda))t} d\lambda , \tag{2.80}$$

where we assumed that the contour is already sufficiently long so that it samples all possible stretching rates of the different flow regions. For large t the dominant contribution in (2.80) comes from a particular Lyapunov exponent λ^* for which the exponent has a minimum

$$\lim_{t \to \infty} l(t) \sim e^{\gamma t} \text{ where } \gamma = \max_{\lambda}[\lambda - G(\lambda)]. \tag{2.81}$$

In the dynamical systems terminology the exponent γ is the topological entropy of the advection dynamics, that is a measure of the complexity of the dynamics defined as the growth rate of the number of distinguishable orbits when the spatial resolution is increased. It can be easily shown that the topological entropy is always larger than the leading asymptotic Lyapunov exponent λ^∞ (Fig. 2.12). Thus, the growth of the contour length is dominated by a small proportion of non-typical fluid elements experiencing a larger than average stretching rate. When the non-uniformity of the flow is not too strong the entropy function can be approximated by a parabola, leading to the

Gaussian in (2.76), and the maximum is at $\lambda^* = \lambda^\infty + \Delta$ that gives a contour growth rate

$$\gamma = \lambda^* - G(\lambda^*) = \lambda^\infty + \frac{\Delta}{2}. \qquad (2.82)$$

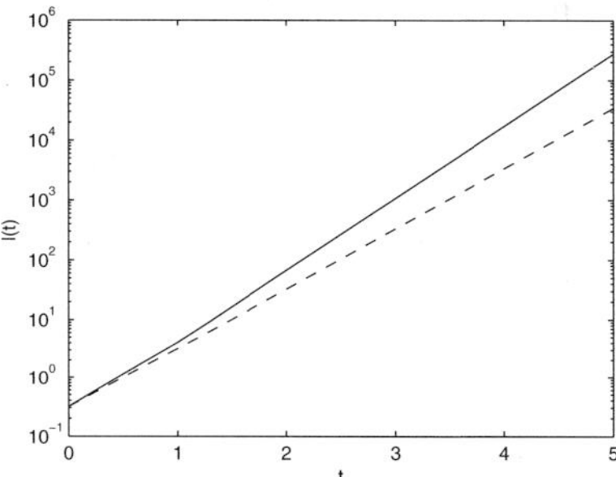

Figure 2.12: Length of the contour shown in Fig. 2.10 as a function of time (solid line), compared to the growth rate predicted by the Lyapunov exponent $l_0 \exp(\lambda t)$ (dashed line).

2.6 Chaotic advection in open flows

A special case of chaotic advection occurs in open flows in which the time-dependence of the flow is restricted to a bounded region (Tél et al., 2005). This kind of flow structure with an unsteady mixing region and simple time-independent inflow and outflow regions is typical for example in stirred reactors or in a flow formed in the wake of an obstacle. A well known example is the von Kármán vortex street behind a cylinder at moderate Reynolds numbers (Jung et al., 1993; Ziemniak et al., 1994), where around the cylinder the flow is time-periodic, but at some distance from it upstream or downstream the velocity field is time independent.

Since the time-dependence of the velocity field is restricted to a finite region the complex chaotic orbits are also limited to this region. Advection in such flows is a chaotic scattering process (Ott and Tél, 1993; Ott, 1993) in which fluid elements approach the mixing zone along the inflow streamlines, they follow chaotic trajectories inside the unsteady mixing zone and eventually leave on simple outflow trajectories. The main feature of chaotic scattering is that slightly different inflow trajectories can lead to very different escape times. The escape time is an irregular function of the inflow coordinate with a set of isolated singularities as shown in Fig. 2.13 for a flow to be defined later (Eq. (2.86)). This type of chaotic scattering has been studied in various dynamical systems and is typically associated with the existence of a bounded non-attracting chaotic invariant set in the phase space. The sensitivity of the escape time is a manifestation of the chaotic nature of the advection within the mixing zone.

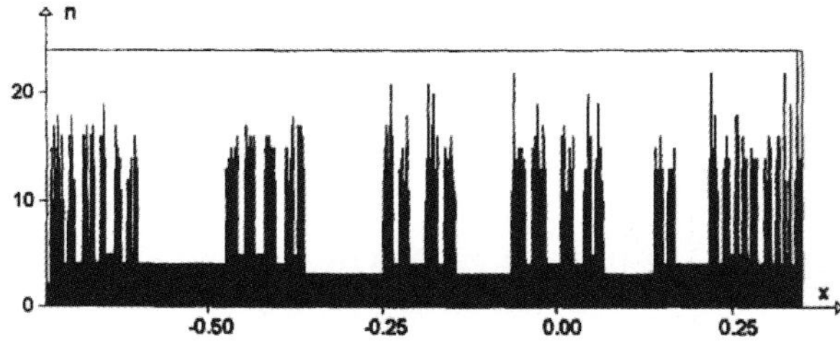

Figure 2.13: Escape times as a function of initial location, showing singularities on a fractal set for an ensemble of particles released on a line segment in the vortex-sink flow (from Karolyi and Tel (1997)).

The non-attracting chaotic set responsible for the chaotic scattering is formed by the set of bounded chaotic trajectories that do not leave the mixing zone neither forward nor backward in time. In the case of a time-periodic open flow there are infinitely many unstable (hyperbolic) periodic orbits within the mixing zone with periods of integer multiples of the flow period. The closure of this set, i.e. the collection of all the periodic orbits together with other orbits that connect them by bouncing indefinitely between these unstable periodic orbits form a hyperbolic invariant set which is called the *chaotic saddle.* Such set of bounded chaotic orbits may also exist in the mixing zone of an aperiodic open flow with the difference that in this case the set does not contain any periodic orbits and the spatial structure of the set changes aperiodically in time following the flow.

Just like an isolated saddle point, the chaotic saddle has stable and unstable manifolds. The stable manifold of the chaotic saddle is a

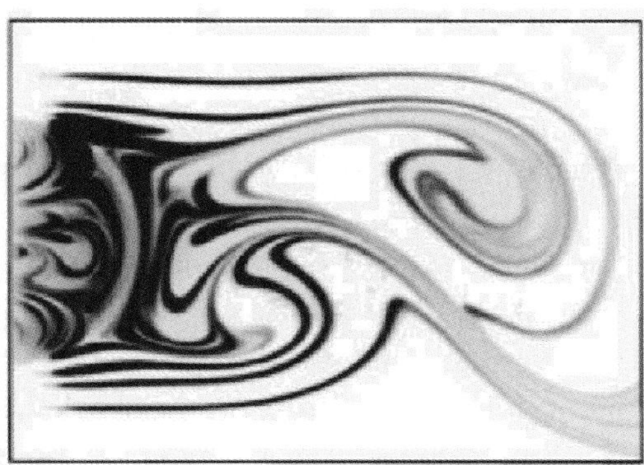

Figure 2.14: Experimental visualization of the unstable manifold in an open flow as the streakline pattern produced by dye injected in the wake of a cylinder (Sommerer et al., 1996).

set that contains all the orbits that lead to the saddle from the inflow region, i.e. all trajectories that never reach the outflow, hence their escape time is infinite. Trajectories from the inflow region that come close to the saddle are trapped around it for long time. The unstable manifold of the chaotic set contains all the orbits that lead to the chaotic saddle from the outflow region in the time-reversed dynamics. The exit trajectories for the orbits with very long escape times, i.e. those that come close to the saddle, will leave the mixing zone along trajectories that are in the close vicinity of the unstable manifold. Therefore a droplet of dye injected into the mixing region, after some transient time, will trace out the unstable manifold making it visible in experiments, as shown in Fig. 2.14.

The chaotic saddle and its manifolds are also sets of zero measure with fractal structure. The set of points, seen in Fig. 2.13 corresponding to inflow coordinates with very large, singular, escape times, typically form also a fractal set determined by the intersection of the saddle's stable manifold and the line containing the initial conditions. There is a connection between the dimension of the chaotic saddle and the dimensions of its manifolds. The trajectories on the chaotic saddle have a set of Lyapunov exponents whose number is equal to the dimension of the full space, d. The sum of the Lyapunov exponents is zero due to incompressibility and chaotic dynamics implies that at least one of them is positive. The fractal dimensions of the stable and unstable manifolds can be written as $D_f^{s,u} = n_{s,u} + \alpha_{s,u}$ where n_s and n_u are the number of negative and positive Lyapunov exponents, i.e. the number of attracting and expanding directions of the chaotic dynamics. These satisfy $n_s + n_u = d$ assuming that none of the exponents is zero. The constants α_s and α_u are positive and generally non-integers that characterize the highly folded geometric structure of the manifolds. Since the chaotic saddle is at the intersection of the stable and unstable manifolds its dimension can be obtained as the dimension of the intersection of two fractal objects, that in a d-dimensional space is $D_f = \min(d, D_{f,1} + D_{f,2} - d)$. For the dimension of the saddle this gives $D_f^{saddle} = \alpha_s + \alpha_u$ so that $0 < D_f^{saddle} < d$. For two-dimensional chaotic flows $n_s = n_u = 1$ and since the two Lyapunov exponents of opposite sign are equal in magnitude the dimensions of the stable and unstable manifolds are also equal and $D_f^{saddle} = 2\alpha$.

In addition to the Lyapunov exponent, that is a measure of chaotic advection inside the mixing zone, the transport in open flows has another characteristic timescale associated to the escape rate of fluid elements. This can be defined from the distribution of escape times τ, that for large values (in hyperbolic systems) has an exponential form

$$P(\tau) \sim e^{-\kappa\tau} \ . \tag{2.83}$$

This defines κ as the *escape rate*. For example, if an ensemble of particles is injected in the mixing zone the number of particles that have not escaped after a certain time t decreases exponentially as

$$N(t) = N(0)\mathrm{Prob}(\tau > t) = N(0) \int_t^\infty P(\tau)d\tau = N(0)e^{-\kappa t}. \tag{2.84}$$

Note that this is very different from a steady open flow with no chaotic advection where the escape time cannot be arbitrarily large. The exponential distribution is a consequence of the chaotic dynamics, since over the characteristic timescale of the flow (T) advection mixes up and redistributes the fluid elements within the mixing zone. Therefore roughly the same proportion $\approx \kappa T$ of the remaining particles escape after each time T. This leads to the exponential decay after iterating many periods T. If the non-escaping set also has a non-hyperbolic component (e.g. due to a boundary layer around a solid surface, or to KAM tori), then the exponential escape of the particles at long times is replaced by a slower algebraic decay $N(t) \sim t^{-\sigma}$ (Lau et al., 1991).

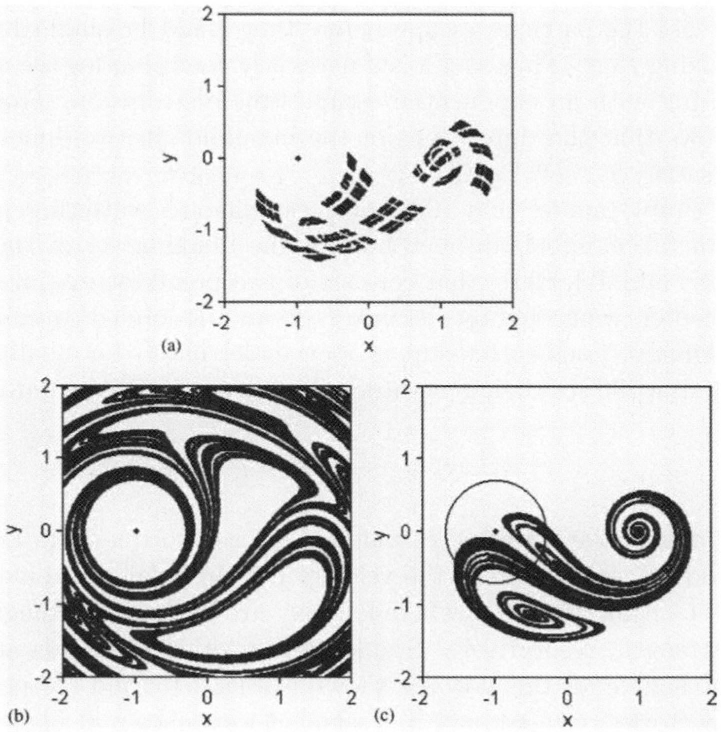

Figure 2.15: Instantaneous view of the chaotic saddle (a), stable (b) and unstable (c) manifolds for the advection dynamics in the blinking vortex-sink flow. The vortex-sinks are at $(x, y) = (-1, 0)$ and $(0, 1)$ (from Karolyi and Tel (1997)).

The dynamical and geometrical properties of the advection in open flows are related through the generalization of the Kaplan-Yorke formula that gives the fractal dimension of the manifolds as a function of Lyapunov exponents and escape rate. In the case of two-dimensional flows this is (Ott, 1993)

$$D_f^{u,s} = 1 + \alpha = 2 - \frac{\kappa}{\lambda}. \tag{2.85}$$

Note that the chaotic scattering can only take place when the typical lifetime of the transient chaotic advection, $1/\kappa$, is longer than the rate of separation of chaotic trajectories on the chaotic saddle, $1/\lambda$. Otherwise the particles escape before they could be shuffled within the mixing zone. Thus $\kappa < \lambda$ is a necessary condition for the chaotic scattering with an exponential escape time distribution. From this it follows that the dimensions of the manifolds in two-dimensional flows satisfy $1 < D_f^{u,s} < 2$.

A simple model that illustrates the chaotic scattering process and fractal manifolds in open flows is the blinking vortex-sink flow (Karolyi and Tel, 1997) that consists of two point vortex-sinks, with their centers separated by distance L, in an unbounded two-dimensional domain. Each vortex-sink is opened alternately for a half period $T/2$ and, while active, it generates a flow given by the streamfunction

$$\psi(r,\varphi) = -\frac{\Gamma}{2\pi}\ln r - \frac{C}{2\pi}\varphi. \tag{2.86}$$

r and ϕ are polar coordinates centered at each vortex-sink. The corresponding components of the velocity field in polar coordinates are $v_r = -C/r$ and $v_\varphi = \Gamma/r$ where Γ and C are the vortex strength and sink strength, respectively. Stable and unstable manifolds and the chaotic saddle for this flow are shown in Fig. 2.15, and escape times of trajectories released on a line segment were shown in Fig. 2.13.

2.7 Chaotic advection and diffusion

Although advection plays a key role in bringing together fluid regions of different properties it cannot produce a spatially uniform concentration field without the help of small scale diffusive transport. In this section we discuss the combined effects of chaotic advection and diffusion. Consider an initial-value problem starting with a spatially non-uniform concentration field, e.g. a small localized patch superimposed on an otherwise uniform background. In a closed flow on a bounded domain the concentration field evolves towards a spatially uniform asymptotic state.

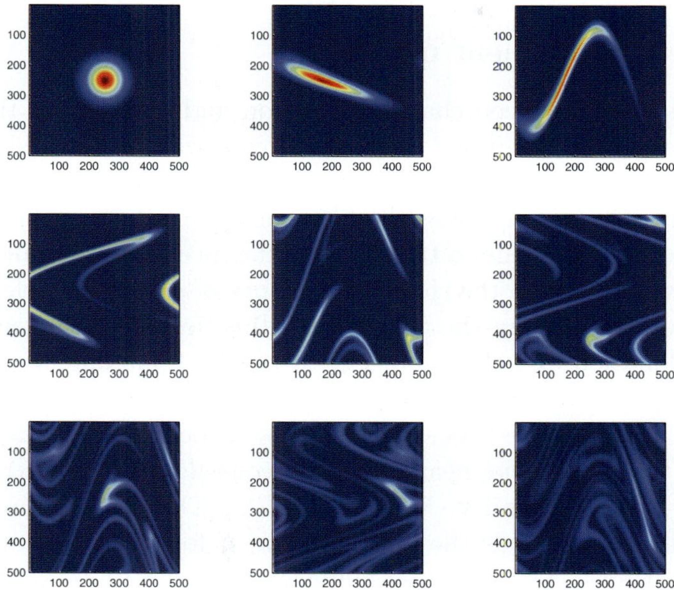

Figure 2.16: The evolution of a localized patch advected by the sine-flow (2.66) in the presence of diffusion. Time goes from left to right and top to bottom.

Snapshots of a concentration field obtained numerically by solving the advection-diffusion equation with the velocity field (2.66) are shown in Fig. 2.16.

We can identify three different stages of mixing. First there is a purely advective regime in which diffusion is negligible and the concentration iso-contours behave as material lines advected by the flow as in Fig. 2.10. As advection generates smaller structures with larger gradients diffusion becomes important. In this intermediate regime the characteristic feature of the concentration field is the presence of thin filaments, that result from a balance between the stretching by the chaotic advection and spreading due to diffusion. Later the filaments densely fill the domain and overlap producing a complex concentration field with multiple scales, whose contrast gradually fades away into a uniform background.

2.7.1   The filament model

The interplay between chaotic advection and diffusion in the intermediate regime of mixing can be well captured by a simple one-dimensional model, first introduced by Ranz (1979), that describes the concentration profile across the filaments formed in two-dimensional chaotic flows. Consider a Lagrangian reference frame co-moving with the flow centered on the chaotic trajectory of a fluid particle with its x axis oriented along the contracting direction $M\mathbf{v}_2$ tangent to the stable foliation and the y tangent to the unstable one (we are thus assuming hyperbolicity). Close to the point, the linearized flow reads $\mathbf{v} = (-\lambda x, \lambda y)$. The concentration C is stretched along the y direction so that it becomes nearly homogeneous along it: $C(x, y) \approx C(x)$, $\nabla^2 C \approx \partial^2 C/\partial x^2$, and $\mathbf{v} \cdot \nabla C \approx v_x \partial C/\partial x$. The flow compresses the concentration towards the y axis so that it forms a filament aligned with that axis. The advection-diffusion equation (2.31) for $C(x)$, in the absence of sources and close to the hyperbolic point becomes:

$$\frac{\partial C}{\partial t} - \lambda x \frac{\partial C}{\partial x} = D \frac{\partial^2 C}{\partial x^2}, \qquad (2.87)$$

where D is the molecular diffusion coefficient. Since this equation represents the filament profile in a Lagrangian reference frame, the appropriate strain rate in (2.87) is the Lyapunov exponent: $\lambda = |\lambda_2|$. Assuming that the background concentration is zero and the distance between other possible filaments in the flow is sufficiently large, the boundary conditions can be chosen as $C(x \to \pm\infty) = 0$.

Equation (2.87) represents the competition between the convergent component of the advecting flow, that compresses the concentration towards the origin, and diffusive dispersion. Since this simple model equation aims only to describe the dominant processes, we neglect secondary effects like the fluctuations of the stretching rate and curvature effects due to the folding of the filament. We also neglect for the moment any time-dependence in λ, although this will be relaxed later. Note that Eq. (2.87), when considered as one-dimensional, does not conserve the total concentration ($C_T \equiv \int_{-\infty}^{\infty} dx\, C(x,t)$) but we have instead

$$\dot{C}_T = -\lambda C_T. \qquad (2.88)$$

This is because it represents an open one-dimensional flow in which fluid continuously escapes along the unstable foliation, that is not included in the model (2.87). The loss of fluid along the unstable direction, however, is balanced by the apparent compressibility of the flow along the x direction, $\partial v_x / \partial x = -\lambda$, which is consistent with the loss rate in (2.88).

The asymptotic solution of (2.87) for large times is a Gaussian concentration profile with exponentially decaying amplitude

$$C(x,t) = C_0 e^{-\lambda t} e^{-\lambda x^2/(2D)}, \qquad (2.89)$$

representing a filament of width $l_D = (D/\lambda)^{1/2}$. It is characterized by a diffusive scale arising from the balance between diffusion and the convergent advection. The decreasing amplitude is a consequence of the continuous escape of fluid in the one-dimensional description (2.88), that in the full two-dimensional problem is compensated by the increasing length of the filament along the unstable foliation. Since the average width of the growing filament is constant, its area is not conserved as in the case of pure advection, but increases in time as

$$A(t) \approx l_D l(t) \sim \sqrt{\frac{D}{\lambda}} e^{\lambda t}. \qquad (2.90)$$

This also implies that the concentration within the filament should decrease as the initial patch is diluted by the fluid entrained into the filament along the contracting direction.

The general time-dependent solution which is normalizable (i.e. with finite C_T) of Eq. (2.87) can be found in terms of Hermite polynomials as

$$C(x,t) = e^{-\lambda t} e^{-\frac{x^2 \lambda}{2D}} \sum_{n=0}^{\infty} C_n e^{-n\lambda t} H_n\left(x\sqrt{\lambda/2D}\right). \qquad (2.91)$$

This solution shows explicitly that the basic decaying filament solution (2.89) is the dominant structure approached at long times. Figure 2.17 shows an example of time-dependent evolution from a step-like initial condition leading to the asymptotic state (2.89).

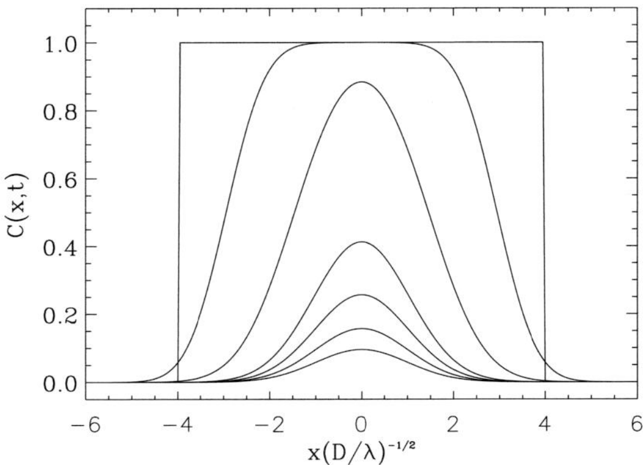

Figure 2.17: Time-dependent solution of Eq. (2.87) with a step-like initial condition. Times shown are $\lambda t = 0, 0.03, 1, 2, 2.5, 3$, and 3.5, in order of decreasing maximum height. Solution (2.89) is approached after $\lambda t \gtrsim 1$.

Another description of the dynamics of (2.87) can be extracted from the observation that there is a change of variables that eliminates the advection term in (2.87). This is done (Ranz, 1979), even for the case of a time-dependent $\lambda = \lambda(t)$, by replacing x and t by new spatial and temporal coordinates ξ and τ (a stretched coordinate and a warped time) given by

$$\xi = \xi(x,t) \equiv \frac{x}{s(t)}$$

$$\tau = \tau(t) \equiv D \int_0^t \frac{du}{s(u)^2} , \tag{2.92}$$

where

$$s(t) \equiv s_0 e^{-\int_0^t \lambda(u)du} , \tag{2.93}$$

or

$$\lambda(t) = -\frac{1}{s(t)}\frac{d}{dt}s(t) \ . \tag{2.94}$$

The exponential in (2.93) is the factor by which a material length s_0 oriented initially along x (i.e. always in the direction transverse to the filament) is reduced after a time t, thus becoming $s(t)$. In the new coordinates, the equation for the concentration $\tilde{C}(\xi,\tau)$ defined by $\tilde{C}(\xi(x,t),\tau(t)) = C(x,t)$ reads

$$\frac{\partial \tilde{C}}{\partial \tau} = \frac{\partial^2 \tilde{C}}{\partial \xi^2} \ . \tag{2.95}$$

The advection term has disappeared, thus allowing explicit solutions in specific cases. The solution of the diffusion equation (2.95) for vanishing concentrations at infinity is (2.10 – 2.11):

$$\tilde{C}(\xi,\tau) = \frac{1}{\sqrt{4\pi\tau}} \int d\xi\, e^{-\frac{(\xi-\xi')^2}{4\tau}}\, \tilde{C}(\xi',\tau=0). \tag{2.96}$$

The simplest situation to which the transformation (2.92) can be applied is the case of constant λ:

$$s(t) = s_0 e^{-\lambda t} \ , \ \xi = \frac{x}{s_0}e^{\lambda t} \ , \ \tau = \left(e^{2\lambda t} - 1\right)\frac{D}{2\lambda s_0^2} \ . \tag{2.97}$$

Transforming the solution (2.96) back to $C(x,t) = \tilde{C}(\xi(x,t),\tau(t))$ we have another form of the exact time-dependent solution. For $t \gg \lambda^{-1}$ and finite ξ' we have $(\xi-\xi')^2/\tau \approx \xi^2/\tau \approx 2\lambda x^2/D$ and we recover again (2.89). Thus for any sufficiently localized initial condition, the solution is localized in the transverse direction x, representing a concentration filament.

As a second example we consider the case of a shear flow $v_x(y) = Gy$, $v_y = 0$. It was considered using other methods by Birch et al. (2007) or Thiffeault (2008). An initial concentration patch aligned along the y axis has an initial width s_0. As the patch is advected by the flow and becomes tilted at an angle $\alpha(t)$ with the horizontal axis (see Fig. 2.18), the horizontal width remains invariant in time, but the *transverse* width gets reduced: $s(t) = \sin\alpha(t) = $

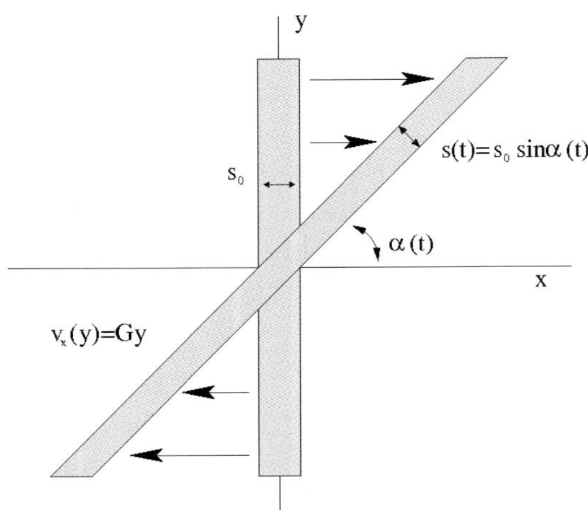

Figure 2.18: Deformation of a filament by a shear flow.

$s_0 \left(1 + (Gt)^2\right)^{-1/2}$. From (2.94), the factor $\lambda(t)$ producing this effect is $\lambda(t) = G^2 t \left(1 + (Gt)^2\right)^{-1}$. We have $\xi = x/s(t)$ and $\tau = Dt/s_0^2 + DG^2 t^3/(3s_0^2)$. At long times the argument of the exponential in (2.96) becomes $\xi^2/4\tau \approx 3x^2/(4Dt)$. Thus the filament still spreads diffusively at long times in the transverse dimension, although with a reduced diffusion coefficient $D/3$. A shear flow is not enough to stop the spreading of a concentration filament. The concentration of the filament at its center decreases as $t^{-3/2}$ so that its total mass decays as t^{-1}. It should be remarked that this behavior is just the final long-time one, at which concentrations are already very small and thus may not be the most relevant in applications. For example, at times before diffusion becomes relevant one finds *shear thinning* by the flow ($s/s_0 \sim G/t$). The full dynamics can be analyzed in detail at all times, as done by Birch et al. (2008) in the context of explaining the formation of horizontal layers of plankton in the sea.

Another example, shown in Fig. 2.19, is an initially straight line rolled up by a point vortex, with flow velocity given, in terms of the

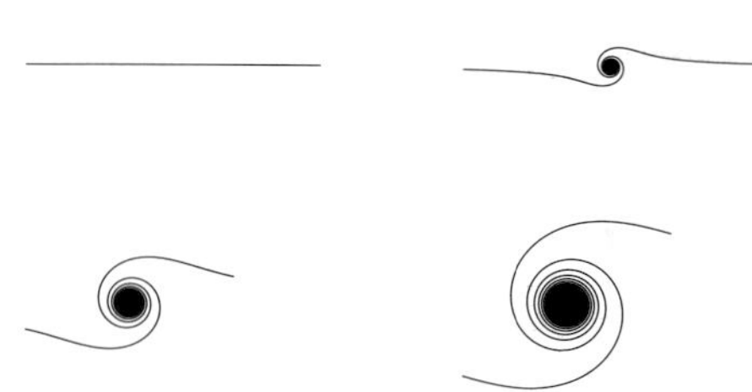

Figure 2.19: A material line rolled up by a point vortex.

angular (ϕ) and the radial (r) coordinates centered at the vortex, by

$$r\dot{\phi} = v_\phi = \frac{\Gamma}{2\pi r} \tag{2.98}$$

$$\dot{r} = v_r = 0. \tag{2.99}$$

After a transient, that depends on the initial condition, the line becomes a spiral given by $\phi = \Gamma t/(2\pi r^2)$. A concentration patch advected by this flow forms a long spiral filament. In order to calculate the width contraction in the direction transverse to the filament, we notice that, for Δr small, we have $\Delta r \approx \pi r^3 \Delta\phi/(\Gamma t)$. We can then estimate the distance between successive turns of the filament, where they are sufficiently close, by setting $\Delta\phi = 2\pi$, that gives $\Delta r \approx 2\pi^2 r^3/(\Gamma t)$. Assimilating this distance contraction with the one experienced by the filament width, and noticing that any given fluid element maintains its radial coordinate constant, we see that the transverse width of a sufficiently thin filament shrinks in time according to $s(t)/s_0 = t_0/t$, where s_0 is the width at time t_0. From (2.94) we have $\lambda(t) = 1/t$. Thus, $\xi = xt/s_0 t_0$, $\tau = Dt^3/(3s_0^2 t_0^2)$, and then from Eq. (2.96) we have again long time diffusive behavior for the filament width despite the Lagrangian contraction of the transverse length. In fact the diffusive behavior at long times is again characterized by the same effective diffusion coefficient, $D/3$, as in the pure shear flow, indicating that locally this is also a shear type flow.

All the above examples have considered isolated filaments. In chaotic flows, where filaments are repeatedly stretched and folded, eventually they will start to overlap. To apply the same formalism for such case we consider an initial condition consisting of an array of filaments with period p, that can be written as a Fourier series $C_0(x) = \sum_{n=-\infty}^{\infty} c_n e^{i2\pi nx/p} = \tilde{C}_0(\xi)$, with

$$c_n = (1/p) \int_{-p/2}^{p/2} dx\, C_0(x) e^{-i2\pi nx/p} \equiv \tilde{c}_n.$$

The solution of (2.95) with such initial condition is (we take $s_0 = p$)

$$\tilde{C}(\xi, \tau) = \sum_{n=-\infty}^{\infty} \tilde{c}_n e^{-4\pi^2 n^2 \tau} e^{i2\pi n\xi} \tag{2.100}$$

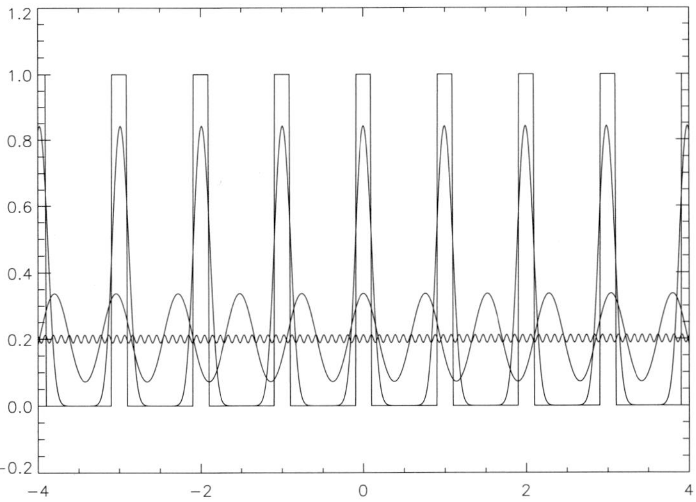

Figure 2.20: The effect of convergent flow and diffusion on a periodic array of filaments, from Eq. (2.87). Configurations with smaller contrast correspond to later times.

By substituting the expressions for $\xi(x, t)$ and $\tau(t)$ we obtain the complete time-dependent solution. As an example we plot in Fig.

2.20 the time evolution of a set of initially sharp filaments ($C_0(x) = 1$ if $|x| < l_0/2$, zero elsewhere in $[-p/2, p/2]$, and repeated periodically in $[-\infty, \infty]$ with period p), for which $c_n = \tilde{c}_n = (\pi n)^{-1} \sin(l_0 \pi n/p)$ if $n \neq 0$ and $c_0 = \tilde{c}_0 = l_0/p$. For the stretching characterizing the flow we have considered the case of a constant λ, so that expressions (2.97) apply. Initially Gaussian-type filaments are formed, that collide, merge and homogenization occurs. Note that the decay becomes superexponential when $\lambda t \gtrsim 1$. This is seen by considering the slowest decaying terms in (2.100), of amplitude $\exp(-4\pi^2 \tau)$, with τ growing exponentially from (2.97). This extremely fast decay is a pathology due to the single value of λ and can be easily understood: The amplitude of a mode with a fixed characteristic wavenumber k decays exponentially under the influence of diffusion as $\exp(-Dk^2 t)$, while the convergent flow increases the wavenumber of the periodic structure exponentially as $k(t) \sim \exp(\lambda t)$. These combined together produce the superexponential decay. Such superfast decay is also found in discrete time models that represent advection by a baker map with a uniform stretching, and similar models (see e.g. Fereday et al. (2002); Thiffeault and Childress (2003); Giona et al. (2005)).

In generic chaotic flows, however, the rate of convergence on the filaments, λ, is not uniform in space and the concentration variance is not simply transferred to higher wavenumber modes at an exponential rate, but the merging of filaments of different sizes leads to a dispersion in wavenumber space that produces a slower, only exponential, decay of the amplitude of inhomogeneities in the concentration field. This regime cannot be properly described in the framework of the simple filament model and the whole concentration field needs to be considered, as described in the next section.

2.7.2 Asymptotic decay in chaotic flows

In the presence of advection and diffusion in a bounded domain the concentration field becomes more and more uniform in space. The distribution can be characterized by its variance, that in the case of chaotic advection decays exponentially in time.

In the simplest case when chaotic advection is generated by a time-independent flow (e.g. in a three dimensional system) the asymptotic behavior can be described by the eigenvalues of the advection-diffusion operator

$$\mathcal{L}^{adv-diff} \equiv \mathbf{v} \cdot \nabla + D\nabla^2. \tag{2.101}$$

Since the average concentration, or total amount, is conserved by advection and diffusion, a zero eigenvalue always exists. The asymptotic decay of the non-uniform component of the concentration field is determined by the second, i.e. the largest non-zero eigenvalue

$$C(\mathbf{x}, t) \simeq \langle C(\mathbf{x}) \rangle + \tilde{C}(\mathbf{x})e^{-\gamma t}, \tag{2.102}$$

where $\tilde{C}(\mathbf{x})$ is the eigenfunction corresponding to the most slowly decaying mode with the negative eigenvalue, $-\gamma$.

The situation is similar in the case of time-periodic flows (Liu and Haller, 2004), where the eigenmodes are also time-dependent. The same advection and diffusion process is applied to the concentration field during each period T, that is given by the linear propagator

$$\mathcal{P}_T^{adv-diff} C(\mathbf{x}, t) = C(\mathbf{x}, t + T). \tag{2.103}$$

Assuming that this operator has a discrete spectrum of time-periodic eigenfunctions $\{C_i\}$ that satisfy

$$\mathcal{P}_T^{adv-diff} C_i = e^{p_i T} C_i , \tag{2.104}$$

where p_i are a set of Floquet exponents, we have $p_0 = 0, C_0(\mathbf{x}) =$ constant, and $0 > \mathrm{Re}\, p_1 > \mathrm{Re}\, p_2 > \cdots$. At long times the spatial structure is dominated by the eigenmode with the slowest decay rate, such that:

$$C(\mathbf{x}, t) \simeq \langle C(\mathbf{x}) \rangle + e^{p_1 t} C_1(\mathbf{x}, t \bmod T). \tag{2.105}$$

Thus, after a transient time the spatial pattern becomes time-periodic (or quasiperiodic when p_1 is complex) with an exponentially decaying amplitude. Such surprisingly simple periodic behavior in time, in sharp contrast with the complex irregular spatial structure of the concentration field, has been first observed in numerical simulations by Pierrehumbert (1994) and later was also confirmed experimentally by Rothstein et al. (1999). In analogy with the rich multiscale spatial structure of the so called *strange attractors* of dissipative chaotic dynamical systems, the eigenmodes of the advection-diffusion operator have been termed *strange eigenmodes* (Pierrehumbert, 1994).

In partially chaotic flows that contain KAM tori and elliptic islands, there is only slow diffusive transport across the invariant tori. Therefore the global decay rate vanishes in the Pe $\to \infty$ (i.e. $D \to 0$) limit. It was found numerically that the decay rate follows a power law $\gamma \sim \text{Pe}^{-\alpha}$, with an exponent in the range $0 < \alpha \leq 1$, while the eigenmodes in such systems are mostly localized in the non-mixing region of the flow (Giona et al., 2004; Pikovsky and Popovych, 2003).

The exponential asymptotic decay also holds for temporally random (i.e. with finite correlation time) spatially smooth flows. In this case the corresponding eigenmode is a stationary random function (Fig. 2.21) and the decay process can be characterized by a set of exponents γ_n, associated to the moments of the concentration field, defined by

$$\langle [C(\mathbf{x}, t) - \langle C(\mathbf{x}, t)\rangle]^n \rangle \sim e^{-\gamma_n t} \qquad (2.106)$$

where $\langle ... \rangle$ represents a spatial average. In flows with no transport barriers the exponents γ_n have a non-zero value in the Pe $\to \infty$ limit, that are thus independent of the diffusivity and represent an intrinsic property of the advecting flow. (Exceptions to this are flows with no-slip boundary conditions in a closed domain where at long times there are concentration fluctuations trapped in the boundary layer leading to slower decay with $\gamma \sim \text{Pe}^{-1/2}$ (Chertkov and Lebedev, 2003).)

The question of what controls the asymptotic decay rate and how is it related to characteristic properties of the velocity field has been an area of active research recently, and uncovered the existence of two possible mechanisms leading to different estimates of the decay rate. Each of these can be dominant depending on the particular system. One theoretical approach focuses on the small scale structure of the concentration field, and relates it to the Lagrangian stretching histories encountered along the trajectories of the fluid parcels. This leads to an estimate of the decay rate based on the distribution of finite-time Lyapunov exponents of the chaotic advection. Details of this type of description can be found in Antonsen et al. (1996); Balkovsky and Fouxon (1999); Thiffeault (2008). Here we give a simplified version of this approach in term of the filament model based

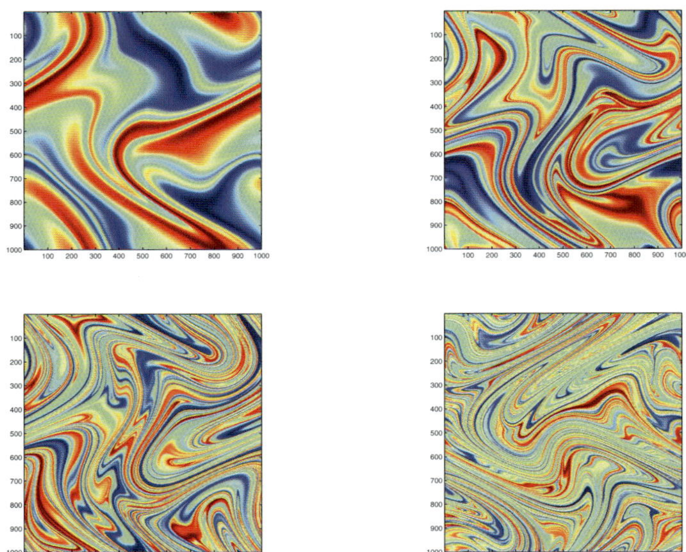

Figure 2.21: Development of the *strange eigenmode* in the sinusoidal shear flow (Eq. (2.66)) with random phase and $\alpha = 0.6$. Time increases from left to right and top to bottom. Since the amplitude of the concentration fluctuations decreases in time the grayscale has been rescaled for each snapshot to cover the range between minimum and maximum concentrations.

on Sukhatme and Pierrehumbert (2002) that leads to essentially the same result.

Assume that the concentration field can be described as a superposition of concentration fluctuations of filamental form, and take into account that each filament experiences a different stretching rate, λ, that satisfies the distribution of finite-time Lyapunov exponents (Eq. (2.75)). Neglecting the differences in the width of the decaying filaments, the moments of the fluctuations of the concentration field can be written as a sum of decaying amplitudes weighted by the distribution of the stretching rates

$$\langle |C - \langle C \rangle|^n \rangle \sim \int e^{-\lambda n t} e^{-G(\lambda)t} d\lambda = \int e^{-t(\lambda n + G(\lambda))} d\lambda. \qquad (2.107)$$

From here the decay rates are obtained by minimizing with respect to λ the coefficient of $-t$ in the exponential, that gives $\gamma_n = n\lambda_n^* + G(\lambda_n^*)$, where λ_n^* is defined by the condition $G(\lambda_n^*)' = -n$. For small deviations from the asymptotic Lyapunov exponent, that is appropriate for n not too large, we can use the approximation $G(\lambda) = (\lambda - \lambda^\infty)^2/(2\Delta)$ to see that the dominant contributions come from values of λ smaller than the asymptotic one, and obtain

$$\gamma_n = n\left(\lambda^\infty - \frac{n\Delta}{2}\right). \tag{2.108}$$

Thus the set of decay rates of the moments is a nonlinear function of the order, n, as proposed by Balkovsky and Fouxon (1999). We remind that the expression (2.108) is not reliable for large n.

Subsequently, Fereday et al. (2002), Sukhatme and Pierrehumbert (2002), and Fereday and Haynes (2004) found that the decay rates predicted from the distribution of Lyapunov exponents do not always agree with the results of numerical simulations, and suggested that they cannot be obtained solely from the statistics of local Lagrangian stretching rates since they also depend on the large scale global structure of the flow. It was also shown that the decay exponent may also depend on the size of the domain, while the distribution of the stretching rates remains the same.

The contradiction with the previous result can be resolved by realizing that the decay is limited either by the small scale structure of the concentration field, or it is controlled by the large scale non-uniformity of the flow (Tsang et al., 2005). Haynes and Vanneste (2005) have shown that the advection diffusion operator has a continuous spectrum in the $D \to 0$ limit whose lower limit coincides with the decay rate predicted from the distribution of the Lyapunov exponents. However, an additional discrete spectrum could also exist and in some cases the most slowly decaying mode may belong to the discrete spectrum. In this case the decay rate is smaller than the one based on the statistics of stretching rates and will depend on the size and shape of the domain. Typically, the decay rate due to the continuous spectrum dominates when the size of the domain is comparable to the characteristic scale of the velocity field, while the global modes dominate on larger domains as the long range transport is not captured by the local Lagrangian theory. In particular, when the domain size (L) is much larger than the characteristic scale of the flow, the advection-diffusion problem can be approximated by a diffusion process with a large-scale effective diffusivity D_{eff} (e.g., Sect. 2.2.2) and the decay rate is $\gamma \sim D_{eff}/L^2$.

2.8 Mixing in turbulent flows

2.8.1 Relative dispersion in turbulence

In the previous sections we considered flows with a smooth spatial structure in which the relative dispersion of fluid trajectories is exponential in time and can be characterized by a single timescale, the inverse of the Lyapunov exponent. This is also valid for two-dimensional turbulent flows that have a smooth velocity field in the small-scale enstrophy cascade range (Bennett, 1984). A similar behavior occurs in any dimension at scales below the Kolmogorov scale (the so-called Batchelor or viscous-convective range, see below). In the inertial range of fully developed three-dimensional turbulence, however, the velocity field has a broad range of timescales and they all contribute to the relative dispersion of particle trajectories and affect the transport properties of the flow.

The relative motion of particle pairs in three-dimensional homogeneous turbulence was first described by L.F. Richardson based on experimental data on atmospheric dispersion. Well before the theoretical results of Kolmogorov on turbulent flows he suggested that the average distance between pairs of particles grows according to a power law of the form (Richardson, 1926)

$$\langle (\mathbf{r}_2(t) - \mathbf{r}_1(t))^2 \rangle \sim t^3. \tag{2.109}$$

Using the result of Kolmogorov for the longitudinal velocity structure function

$$\frac{\langle [\mathbf{v}(\mathbf{r}_2) - \mathbf{v}(\mathbf{r}_1)](\mathbf{r}_2 - \mathbf{r}_1) \rangle}{|\mathbf{r}_2 - \mathbf{r}_1|} = C\epsilon^{1/3}|\mathbf{r}_2 - \mathbf{r}_1|^{1/3} \tag{2.110}$$

we can write the growth rate of the square of the separation as

$$\frac{d}{dt}\langle(\mathbf{r}_2(t) - \mathbf{r}_1(t))^2\rangle = 2\langle(\mathbf{r}_2(t) - \mathbf{r}_1(t))(\mathbf{v}(\mathbf{r}_2) - \mathbf{v}(\mathbf{r}_1))\rangle$$

$$= 2C\epsilon^{1/3}|\mathbf{r}_2 - \mathbf{r}_1|^{4/3} \qquad (2.111)$$

where C is a constant and ϵ is the energy dissipation rate. Solving this equation for the separation we get

$$\langle|\mathbf{r}_2 - \mathbf{r}_1|^2\rangle = \left(\frac{2}{3}C\right)^3 \epsilon t^3 \qquad (2.112)$$

that confirms the Richardson law. This scaling is valid for separations within the inertial range, $\eta < |\mathbf{r}_2 - \mathbf{r}_1| < L$, where η is the Kolmogorov dissipation scale and L is the integral scale of the turbulent flow (i.e. the distance over which the velocity field becomes uncorrelated). For distances larger than the integral scale the separation grows diffusively as given by Taylor dispersion (see Eq. (2.24)), while below the Kolmogorov scale where the flow is smooth, a regime known as the Batchelor regime (or viscous-convective range, see below), an exponential separation is expected as in chaotic advection.

Note that the Richardson law indicates a faster than diffusive separation and can also be interpreted as a scale dependent diffusion coefficient $D = C\epsilon^{1/3}|\mathbf{r}_2 - \mathbf{r}_1|^{4/3}$. A classic example of such length-scale dependent diffusivity is illustrated in the diagram by Okubo (1971), Fig 2.22, that is based on data from various experiments and observations on horizontal dispersion in oceanic flows. These data can be approximately described by a power law dependence of the form $D_{eff}(l) \sim l^{1.15}$. The deviation from the Richardson law can be seen as a consequence of the more complex structure of oceanic currents.

Finite-size Lyapunov exponent

Clearly, the Richardson dispersion law is qualitatively different from the exponential separation of particles in chaotic advection. The finite-size Lyapunov exponent (FSLE) is a generalization of the Lyapunov exponent introduced to characterize flows with multiple length

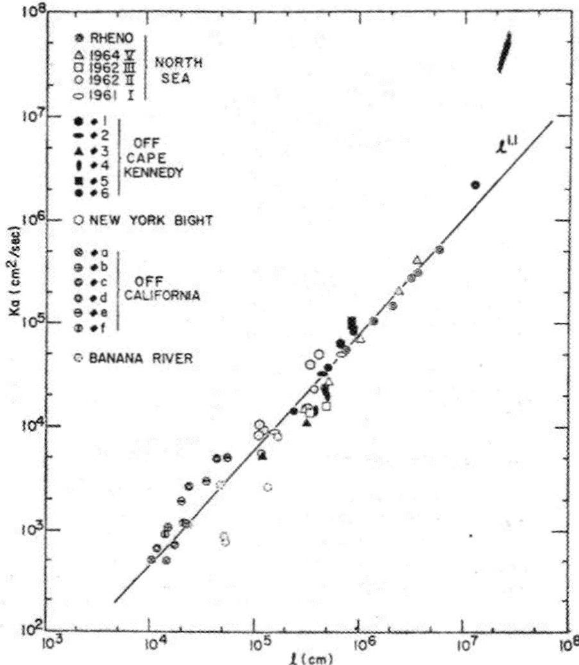

Figure 2.22: Effective diffusivity as a function of length scale for horizontal dispersion in the ocean (Okubo, 1971).

and timescales (Artale et al., 1997; Aurell et al., 1997). It is somewhat similar to the Finite-time Lyapunov exponent (Sect. 2.5.1) which measures dispersion rate averaged over a finite time. Here, the FSLE measures the dispersion rate at a fixed length scale and it is defined as the inverse of the average time needed for an initial distance δ between a pair of advected particles to increase by a certain fixed ratio $r > 1$ (e.g. $r = 2$ for doubling times)

$$\lambda(\delta) = \frac{\ln r}{\langle T_r(\delta) \rangle}. \qquad (2.113)$$

In the $\delta \to 0$ limit this is equivalent to the standard Lyapunov exponent, but when δ is non-zero it represents the separation rate representative to that particular length scale in the flow, assuming that r is not too large. The FSLE can also be used to define a scale-dependent diffusion coefficient as $D(\delta) = \delta^2 \lambda(\delta)$. Thus when the dispersion is a diffusion-type process over a certain range of length scales, the FSLE has a form $\lambda(\delta) \sim \delta^{-2}$, corresponding to a constant scale-independent effective diffusivity.

For fully developed three-dimensional turbulence one can identify different regimes. Below the Kolmogorov scale the velocity field is smooth and chaotic advection dominates, therefore the FSLE is the same as the standard Lyapunov exponent and is independent of δ. When δ is within the inertial range, we have $\lambda(\delta) \sim \delta^{-2/3}$. If the size of the domain is much larger than the integral scale this is followed by a diffusive regime where $\lambda(\delta) \sim \delta^{-2}$ for $\delta \gg L$. Thus the FSLE provides a well defined measure of the transport timescales associated to different length scales that can be applied to any complex velocity field. The form of the function $\lambda(\delta)$ can be used to identify different characteristic flow regimes, and the FSLE also exists in crossover regions that are not well characterized by standard measures like the Lyapunov exponent or the effective diffusivity.

The application of FSLE was demonstrated in the analysis of experimental (Boffetta et al., 2000) and geophysical flows (Lacorata et al., 2001). Instead of calculating the average over many particle pairs in Eq. (2.113), one can measure the separation rates as a function of the initial position of the trajectories. This type of analysis has been used for visualizing the location of transport barriers and coherent structures in geophysical flows (Boffetta et al., 2001; d'Ovidio et al., 2004, 2009). Note that the distribution of finite time Lyapunov (Lapeyre, 2002) exponents also produces similar structures in some range of parameters, namely small initial separation and r (for FSLE) and large time (for FTLE).

2.8.2   Passive scalar in turbulent flows

The theory of mixing of a passive scalar concentration field subject to advection and diffusion in a high Reynolds number turbulent flow is based on the works of Obukhov (1949) and Corrsin (1951). Consider a statistically stationary state with a large-scale source of scalar fluctuations in the case when both Pe and Re are large. The concentration field is governed by the equation

$$\frac{\partial C}{\partial t} + \mathbf{v} \cdot \nabla C = S(\mathbf{x}) + D\nabla^2 C. \tag{2.114}$$

A stationary state with constant average concentration is obtained when S is a source/sink term with zero mean. In this case the fluctuating component of the scalar field $\theta(\mathbf{x}, t) \equiv C(\mathbf{x}, \mathbf{t}) - \langle C \rangle$ also satisfies the above equation and can be characterized by the variance $\langle \theta^2 \rangle$. To obtain an equation for the variance we multiply the equation by θ and integrate over the whole domain

$$\frac{1}{2}\frac{d\langle \theta^2 \rangle}{dt} + \langle \theta \mathbf{v} \cdot \nabla \theta \rangle = \langle S\theta \rangle + D\langle \theta \nabla^2 \theta \rangle. \tag{2.115}$$

Rewriting the advective term as

$$\theta \mathbf{v} \cdot \nabla \theta = \nabla \cdot (\theta^2 \mathbf{v}) - \theta^2 \nabla \cdot \mathbf{v} - \theta \mathbf{v} \cdot \nabla \theta \tag{2.116}$$

the second term on the right is zero due to incompressibility while the first term vanishes after the spatial integration (assuming periodic or no-slip boundary condition). Similarly rewriting the diffusion term gives

$$\frac{1}{2}\frac{d\langle \theta^2 \rangle}{dt} = \langle S\theta \rangle - D\langle (\nabla \theta)^2 \rangle \tag{2.117}$$

where the first term on the right is due to the forcing that injects concentration fluctuations at large scale and is balanced by the second term, arising from diffusive dissipation acting at small scales. Although the advection term does not appear explicitly in this equation, advection influences the distribution of gradients and has an important role in creating the cascade that transports the scalar variance without loss to smaller scales so that finally it can be dissipated by diffusion.

In analogy with Kolmogorov's assumption of the finite energy dissipation rate, Obukhov and Corrsin assumed that as the diffusivity is reduced to zero the gradients increase in such a way that the rate of dissipation of the scalar variance has a finite non-zero limit

$$\lim_{D \to 0} D\langle (\nabla \theta)^2 \rangle = \epsilon_\theta. \tag{2.118}$$

Thus analogously to the inertial scale of turbulence, the statistical properties of the scalar fluctuations in the inertial-convective range, i.e. in a range of scales below the forcing scale where both diffusion and viscosity are negligible, can only depend on the dissipation rate ϵ_θ, the energy dissipation rate ϵ, and on the length scale. Thus the only dimensionally correct form of the second order scalar structure function of the concentration fluctuations is

$$\langle(\theta(\mathbf{r}+\boldsymbol{\delta})-\theta(\mathbf{r}))^2\rangle = K_\theta \epsilon_\theta \epsilon^{-1/3}\delta^{2/3}, \tag{2.119}$$

where K_θ is a dimensionless constant, and the corresponding power spectrum is

$$E_\theta(k) = K'_\theta \epsilon^{-1/3}\epsilon_\theta k^{-5/3}. \tag{2.120}$$

The range of validity of these scalings is limited at small scales by either a diffusive length scale, l_D, where diffusion becomes important, or the Kolmogorov scale η where viscosity becomes dominant and changes the characteristics of the velocity field. When the Kolmogorov cascade extends beyond the inertial-convective range, i.e. $\eta < l_D$ (see Fig. 2.23) the diffusive scale can be defined as the length scale where the timescale of diffusion, l_D^2/D, is comparable to the characteristic time of the eddies. In the Kolmogorov cascade the flow timescales follow the scaling $\tau(\delta) \sim \epsilon^{-1/3}\delta^{2/3}$, thus for the diffusive scale we obtain

$$l_D \sim \epsilon^{-1/4}D^{3/4} = \eta\left(\frac{D}{\nu}\right)^{3/4}. \tag{2.121}$$

This is consistent with our assumption that $l_D > \eta$ only when $\nu/D < 1$. The ratio of the diffusivity of momentum and of the diffusivity of concentration defines the Schmidt number $\mathrm{Sc} = \nu/D$. This description also applies to temperature fluctuations in turbulent flows, when the temperature field can be approximated as a passive scalar. In that case the diffusion coefficient is replaced by the heat conduction coefficient κ and the non-dimensional ratio $Pr = \nu/\kappa$ is known as the Prandtl number.

As in the case of the turbulent velocity field, the Obukhov scaling (2.119) can be generalized to higher order structure functions. Assuming self-similarity of the scalar field would imply that higher order structure functions satisfy $S_\theta^n(\delta) \sim \delta^{n/3}$. However, even for moderately high orders, these scaling exponents do not agree with experiments, and the deviation is much stronger than in the case of the velocity field. It was also shown that the anomalous scaling of the scalar field occurs even in the case of exactly self-similar model flows. The basic mechanisms of this anomalous scaling has been identified recently and anomalous exponents were obtained analytically for certain type of model flows (see the reviews by Shraiman and Siggia (2000) and Falkovich et al. (2001)).

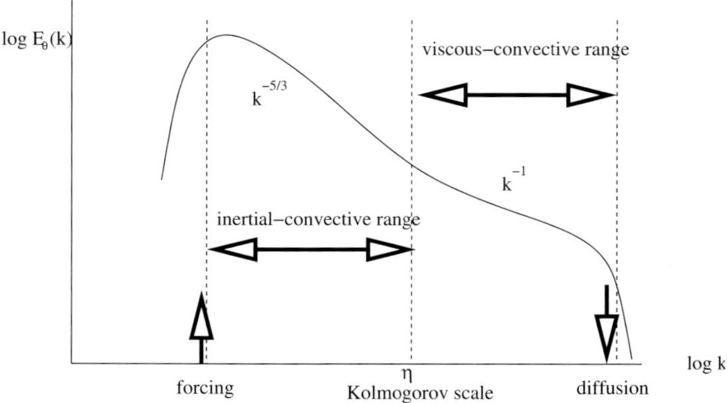

Figure 2.23: Schematic power spectrum of the passive scalar in turbulent flows in the case of large Schmidt number.

The viscous-convective range

When the Schmidt number (or Prandtl number in case of temperature) is larger than unity the Obukhov-Corrsin scaling (2.120) remains valid for scales above the Kolmogorov scale. However, the scalar fluctuations are not yet dissipated at the viscous cut-off, η, and are transferred further to smaller scales by chaotic advection in a spatially smooth unsteady flow. This is the so called Batchelor regime, where the advection is dominated by vortices of a single characteristic size $l \simeq \eta$, with characteristic timescale $\tau_\eta \simeq \epsilon^{-1/2}\nu^{1/2}$. This timescale can also be interpreted as the inverse of the average strain rate, i.e. Lyapunov exponent, experienced by the fluid elements. Assuming again that in the large Pe limit there is a constant flux of scalar variance, independent of the molecular diffusivity, and using the timescale, τ_η and the scalar dissipation rate to construct a dimensionally correct form of the spectrum, leads to (Batchelor, 1959)

$$E_\theta(k) = C\epsilon_\theta \left(\frac{\nu}{\epsilon}\right)^{1/2} k^{-1}. \tag{2.122}$$

This is the Batchelor spectrum, that is valid in the viscous-convective range that extends from the Kolmogorov scale down to the diffusive scale, where the scalar variance is finally dissipated by molecular diffusion. The diffusive scale in this case is the length scale at which the diffusion time l^2/D is comparable to the timescale of advection corresponding to the Kolmogorov scale eddies, that gives

$$l_D = D^{1/2} \left(\frac{\nu}{\epsilon}\right)^{1/4} = \eta Sc^{-1/2}. \tag{2.123}$$

The structure functions in this regime are logarithmic of the form

$$S_\theta^n(\delta) \sim \ln\left(\frac{\delta}{l_D}\right)^{n/2}. \tag{2.124}$$

Note that the inertial range and high Re are not necessary for the existence of this type of Batchelor scaling and it can be also produced by chaotic advection in spatially smooth unsteady large scale flows with relatively small or moderate Reynolds number, or in two-dimensional flows (Jullien et al., 2000; Pierrehumbert, 2000).

2.9   Distribution of inertial particles in flows

Particles of finite size or with a density different from that of the surrounding fluid (e.g. liquid droplets or dust particles suspended in a fluid), due to their inertia and non-vanishing size, have an instantaneous velocity that is somewhat different from the local velocity of the fluid. Therefore such inertial effects can have a significant influence on the distribution of suspended particles. If the Reynolds number based on the size of the particle and its velocity relative to the fluid is small, the flow around the particle can be approximated by the solution of the Stokes equation, and the equation of motion of particles in an incompressible fluid is given by the Maxey-Riley equation (Maxey and Riley, 1983)

$$
\begin{aligned}
m_p \frac{d\mathbf{v}}{dt} \;=\; & m_f \frac{D\mathbf{u}}{Dt} - 6\pi a \nu \rho_f [\mathbf{v} - \mathbf{u}(\mathbf{r},t)] - \frac{m_f}{2}\frac{d}{dt}[\mathbf{v} - \mathbf{u}(\mathbf{r},t)] \\
& - 6a^2 \rho \sqrt{\pi \nu} \int_0^t \frac{ds}{\sqrt{t-s}}\frac{d}{ds}[\mathbf{v} - \mathbf{u}(\mathbf{r},s)] \;,
\end{aligned}
\tag{2.125}
$$

where m_p and a are the mass and radius of the particle, $m_f = \rho_f m_p / \rho_p$ is the mass of the displaced fluid, $\mathbf{v}$ is the velocity of the particle and $\mathbf{u}$ is the fluid velocity. The terms on the right side are the pressure gradient force, the Stokes drag, the added mass force (due to the fluid displaced by the moving particle) and the Basset history force. After non-dimensionalising Eq. (2.125) one obtains two non-dimensional parameters: the added mass factor $\beta = 3m_f/(m_f + 2m_p)$ and the Stokes number $St = (a^2 U)/(3\beta \nu L) = (a/L)^2 \mathrm{Re}/(3\beta)$ that is ratio of the relaxation time towards the local flow velocity (i.e. the Stokes time), and of the characteristic advection time of the flow, L/U. If the particle velocity relative to the fluid is small compared to the characteristic flow velocity and for large Stokes number, or in the case of particles that are much heavier than the fluid ($\beta \ll 1$), the particle motion is described by a simplified system

$$
\frac{d\mathbf{r}}{dt} = \mathbf{v}, \quad \frac{d\mathbf{v}}{dt} = \beta \frac{d}{dt}\mathbf{u}(\mathbf{r},t) - \frac{1}{St}[\mathbf{v} - \mathbf{u}(\mathbf{r},t)].
\tag{2.126}
$$

It is easy to show that the dynamics of the system (2.126) is dissipative. The divergence of the right hand sides with respect to $(\mathbf{r},\mathbf{v})$ is strictly negative and equal to $-d/St$ where d is the space dimension. Then, since the particle velocity $\mathbf{v}$ has a non-zero divergence even in an incompressible velocity field of the fluid, this inertial effect leads to a non-uniform spatial distribution of the particles (Balkovsky et al., 2001a). This was demonstrated numerically by Sigurgeirsson and Stuart (2002) for the case of particles much heavier than the ambient fluid, i.e. when $\beta = 0$. Babiano et al. (2000) have shown that even for neutrally bouyant particles (i.e. for $\beta = 1$) one can see that transport boundaries formed by KAM tori transform into attracting orbits that may cause the trapping of particles in flows with coherent vortex structures.

The non-zero divergence of the particle's velocity field implies the contraction of their phase space volumes. Thus the attractor of the particle trajectories in the phase space spanned by $(\mathbf{r}, \mathbf{v})$ has a fractal dimension $D_f < 2d$, i.e. lower than the full dimension of the phase space. The distribution of the particles in the fluid is given by the projection of the attractor to the d-dimensional physical space. The dependence of the attractor dimension on the parameters β and St was studied by Bec (2003, 2005) for two- and three-dimensional flows generated by a stochastic synthetic turbulence model. Since direct measurement of the fractal dimension is numerically difficult, this was estimated using the Kaplan-Yorke formula (Ott, 1993), that relates the dimension of the attractor to the Lyapunov exponents of the dynamics. When the dimension of the attractor is in the range $d < D < 2d$ the distribution of the particles is space filling. But when the Stokes number is smaller than a critical Stokes number, which is a function of β, the dimension of the attractor is smaller than d, and therefore its projection to the physical space is also a fractal. In this case the particles cluster along filaments with a dynamically changing fractal structure separated by empty regions on all scales (Fig. 2.24). In the case of light particles a third regime of pointwise clustering is also possible in two-dimensional flows when both Lyapunov exponents are negative and all particles converge to the same trajectory.

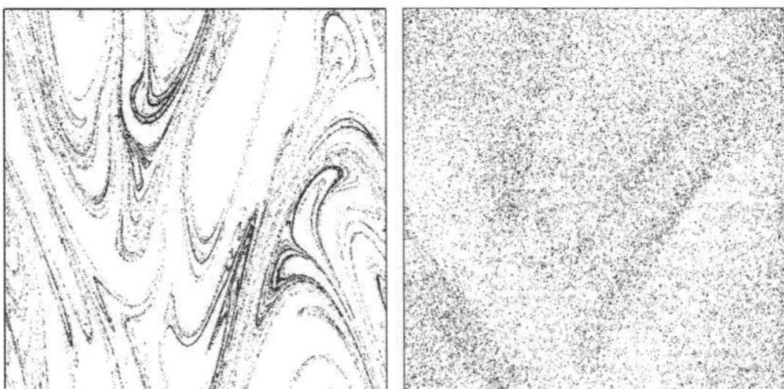

Figure 2.24: Distribution of heavy inertial particles ($\beta = 0$) in a stochastic model flow at two different Stokes numbers $St = 10^{-2}$ (left) and $St = 1$ (right) (Bec, 2003).

Another interesting effect of the particle inertia is that it can transform non-attracting chaotic sets into chaotic attractors. This has been shown by Benczik et al. (2002) who studied the motion of inertial particles in the time-periodic Kármán vortex flow that produces transient chaotic advection of non-inertial particles, while inertial particles are trapped indefinitely in the wake indicating the presence of an attractor.

Chapter 3

Chemical and Ecological Models

After the general description of flows and mixing, we now elaborate on the basics of chemical and biological dynamics, before combining them with the flow processes. There are excellent books describing in detail nonlinear chemical dynamics (Epstein and Pojman, 1998; Scott, 1991; Murray, 1993) and population dynamics (Murray, 1993; Waltman, 1983; Hofbauer and Sigmund, 1988; Farkas, 2001) and this chapter can not be a substitute for them. We just present the building blocks which are being used to model these complex dynamical systems, and summarize the main mechanisms and behavior observed. Our emphasis is on the common features found in rather different chemical or biological situations.

3.1 Chemical dynamics

3.1.1 The Law of Mass Action

The conservation of the atoms involved in chemical reactions is stated in expressions of the type

$$n_a A + n_b B + \cdots \longrightarrow n_x X + n_y Y + \cdots \tag{3.1}$$

involving the stoichiometric coefficients n_a, n_b, n_x, etc. Expression (3.1) means that n_a molecules of A, n_b of B, ... are consumed at each occurrence of the reaction, and that n_x molecules of X, n_y of Y appear. This poses strong constraints on the rate at which the different compounds appear and disappear. In the case of reactions in solution this leads to relationships between the rate of change of the different molar concentrations (denoted by [...]). In particular, (3.1) implies that

$$-\frac{1}{n_a}\frac{d[A]}{dt} = -\frac{1}{n_b}\frac{d[B]}{dt} = \cdots = \frac{1}{n_x}\frac{d[X]}{dt} = \frac{1}{n_y}\frac{d[Y]}{dt} = \cdots \equiv \nu \ . \quad (3.2)$$

The last equality defines the *rate ν* of the chemical reaction. In order to turn the relationships (3.2) into differential equations for the concentrations, the dependence of the rate $\nu = \nu([A],[B],...)$ on the concentrations should be elucidated, which is commonly done experimentally. There are reactions for which the concentration dependence of ν is particularly simple, namely the product of the reactant concentrations raised to some powers, $\nu = k[A]^a[B]^b....$ In such cases, the sum of the exponents $a + b + \cdots$ is called the order of the reaction. In fact, molecular collision arguments (Kreuzer, 1981) imply that if the expression (3.1) really describes the elementary molecular reaction process, then the reaction rate is

$$\nu([A],[B],...) = k[A]^{n_a}[B]^{n_b} \dots \ . \quad (3.3)$$

The reaction constant k, fixing the time scale, may depend on external conditions such as temperature. Expression (3.3) is commonly called the *law of mass action*, although this name is also given to an important relationship among product and reactant concentrations at chemical equilibrium. When the molecular mechanism leading to the reaction (3.1) involves intermediate molecular steps, expression (3.3) can not be used and the exact sequence of intermediate reactions should be established. Or rather an empirical expression for $\nu = \nu([A],[B],...)$ should be used.

In general, if we have M chemicals A_1, $A_2,\ldots$, A_M which can appear both as reactants and as products in N coupled reactions (with reaction constants k_α, $\alpha = 1,\ldots,N$):

$$\sum_{i=1}^{C} r_{i\alpha}A_i \xrightarrow{k_\alpha} \sum_{i=1}^{C} p_{i\alpha}A_i \ , \ \alpha = 1,\ldots,N \quad (3.4)$$

where $r_{i\alpha}$ is the stoichiometric coefficient of the chemical A_i in the reactant side of reaction α, and $p_{i\alpha}$ is the analogous coefficient on the product side (some of them can take zero values), the law of mass action states in this case that

$$\frac{d}{dt}[A_i] = \sum_{\alpha=1}^{N} k_\alpha \left(p_{i\alpha} - r_{i\alpha}\right) \prod_{j=1}^{R_\alpha} [A_j]^{r_{j\alpha}} \qquad (3.5)$$

where the product runs over the R_α reactants involved in reaction α (it can also run over the full set of M chemicals, since the ones which are not reactants in a particular reaction will have the corresponding $r_{j\alpha}$ equal to zero).

In this way we can establish the kinetic equations for the most elementary types of reactions, that will be the building blocks of our further modelling developments. Unless otherwise stated, the chemical reactions written in the following are assumed to be elementary, so that the law of mass action can be applied. In the following we will denote the molar concentrations of the chemical species with the same symbol as the species itself, i.e. $[A] \equiv A$.

We comment on two facts about Eqs. (3.2)-(3.3). First, the coefficients appearing there are correct when the amount of chemicals is expressed in molar concentration or in number of molecules. When using mass or other units the coefficients should be modified accordingly. Second, the use of concentrations taking a continuum of values is justified in most of the standard situations encountered in chemical dynamics, and will be followed in the rest of this book, but the discrete molecular nature shows up when considering reactions involving a small number of molecules, as for example in some biochemical contexts or in genetic regulation (McAdams and Arkin, 1999; Blake et al., 2003). In such cases a stochastic treatment is needed to take into account molecule number fluctuations (Gillespie, 1977; van Kampen, 1992; Gardiner, 2004; Togashi and Kaneko, 2001).

3.1.2 Binary, First-Order, and Zeroth-Order Reactions

As an example of the application of the law of mass action we consider an elementary reaction in which two chemicals are combined:

$$A + B \xrightarrow{k} C \ . \tag{3.6}$$

The concentrations evolve according to

$$\frac{dA}{dt} = \frac{dB}{dt} = -kAB \ , \quad \frac{dC}{dt} = kAB \ . \tag{3.7}$$

The strong symmetry among these three equations imply three conservation laws, two of which are independent:

$$
\begin{aligned}
A(t) + C(t) &= A(0) + C(0) \equiv Q_1 \\
B(t) + C(t) &= B(0) + C(0) \equiv Q_2 \\
A(t) - B(t) &= A(0) - B(0) \equiv Q_3 \ ,
\end{aligned}
\tag{3.8}
$$

so that there is only one independent variable, say A, which satisfies

$$\frac{dA}{dt} = kQ_3 A - kA^2 \ . \tag{3.9}$$

This is an example on how a nonlinear equation for a single chemical can arise by elimination of other substances, and also illustrates that the law of mass action can not be applied to the effective reaction $A \to C$ resulting from (3.6) when eliminating B, since it does not represent the elementary processes (but see later, 3.10). Equation (3.7) or (3.9) with (3.8) describes an evolution in which both reactants A and B decrease in time and C increases, approaching asymptotically a state in which the reactant with the smallest initial concentration disappears.

If in the reaction (3.6) one of the reactants, say B, is kept at constant concentration (so that it is no longer a dynamical variable but a control parameter) or it is in great excess with respect to A (so that its concentration will not decrease substantially during the process) the kinetics will be given by

$$\frac{dA}{dt} = -KA = -\frac{dC}{dt} \ , \tag{3.10}$$

where there is a new constant $K = kB$, and effectively corresponds to applying the law of mass action to $A \to C$. This is an example of *first-order reaction*, i.e. a situation in which the rate is linear in the concentration of one of the reactants. This concentration, thus, decays exponentially in time. Other examples of first-order reactions are isomerization processes, $A \to B$, or decay processes such as radioactive decay, some photochemical processes, or the passage of a chemical to an inert phase, in which a substance disappears or it is removed at a rate proportional to its mass or concentration. These are generically denoted as

$$A \to \emptyset \ . \tag{3.11}$$

The inverse of the reaction (3.6), i.e. the decomposition process

$$C \xrightarrow{k} A + B \tag{3.12}$$

is also a first order reaction, with rate proportional to C:

$$\frac{dA}{dt} = \frac{dB}{dt} = -\frac{dC}{dt} = kC \ . \tag{3.13}$$

Imagine now that the concentration of C is maintained constant. Then, the growth of either A or B occurs at a constant rate, independently of their concentrations. In this case the reaction is said to be of *zeroth order*. This is also the case of the inverse of (3.11), i.e. $\emptyset \xrightarrow{k} A$, that may represent a constant input of a substance from an external source, leading to

$$\frac{dA}{dt} = k \ . \tag{3.14}$$

We finally note that the reaction

$$A_0 \underset{k}{\overset{k}{\rightleftharpoons}} A \tag{3.15}$$

in which both the direct and the reverse reaction occur at the same rate, can be thought as a model for the input and output of reactants in a continuous-flow stirred tank reactor (CSTR), in which external molecules, A_0, enter the reactor and remain there (a state in which they are called A) during a residence time of average duration k^{-1}, before leaving again the system. Note that the idea of full mixing of the reactant within the whole reactor volume before escaping, and the assumption that a fixed portion of the reactor contents leaves it at each time interval, which is needed for a well defined residence time to exist, are the characteristics of the stirring provided by the *chaotic open flows* discussed in Sect. 2.6. Since usually the external concentration A_0 is kept constant, the rate for A in such process is

$$\frac{dA}{dt} = k\,(A_0 - A) \ . \tag{3.16}$$

3.1.3 Autocatalytic and Enzymatic Reactions: The adiabatic elimination

The simplest form of catalysis may be represented by a reaction of the type

$$A + B \to A + C \ . \tag{3.17}$$

Although the stoichiometry of the reaction is effectively equivalent to the simplest $B \to C$, we assume that the rate is influenced by the presence of the catalyst A, and that the elementary molecular form of the reaction is (3.17). An *autocatalytic* reaction is one in which the product (or one of the products) of the reaction is the catalyst itself, e.g. $C = A$, so that

$$A + B \xrightarrow{\;k\;} 2A \tag{3.18}$$

In addition to strictly chemical catalysis processes, Eq. (3.18) may represent other situations such as for example combustion (the combustion of B requires of heat A, but the reaction is exothermic, so that more heat A is produced), or secondary crystallization from a solution (B is the substance in solution which crystalizes in form A if some nuclei of A are already present). If B is kept at constant concentration we have the simplest autocatalytic reaction:

$$A \xrightarrow{\;k\;} 2A \ . \tag{3.19}$$

which implies the kinetic law:

$$\frac{dA}{dt} = kA \ . \tag{3.20}$$

In the absence of any other mechanism, this leads to the exponential unlimited growth of A.

The kinetic equations for (3.18) are

$$\frac{dA}{dt} = kAB \ , \quad \frac{dB}{dt} = -kAB \ . \tag{3.21}$$

Again there is a conservation law

$$A(t) + B(t) = A(0) + B(0) \equiv Q \tag{3.22}$$

so that one can reduce the system to a single equation

$$\frac{dA}{dt} = kQA - kA^2 \ . \tag{3.23}$$

We note that this last equation, the *logistic equation*, is formally identical to (3.9). However, the constants Q_3 and Q have a different expression in terms of the initial conditions (see (3.8) and (3.22)). From the exact solution

$$A(t) = \frac{rA(0)e^{rt}}{r + kA(0)\left(e^{rt} - 1\right)} \tag{3.24}$$

where $r = kQ_3$ for (3.9) and $r = kQ$ for (3.23) we see that in the first case (binary reaction) A will always decrease to a steady state value (which is zero if $Q_3 < 0$), whereas in the second (autocatalytic) case A may increase or decrease approaching always a final value $A = Q > 0$. Examples of different concentration evolutions are presented in Fig. 3.1.

We mention also that the same dynamics of logistic growth and saturation can be obtained by combining the autocatalytic reaction (3.19) with its inverse:

$$A \underset{k_2}{\overset{k_1}{\rightleftharpoons}} 2A \tag{3.25}$$

which leads to

$$\frac{dA}{dt} = k_1 A - k_2 A^2 \ . \tag{3.26}$$

A different kind of catalysis is frequent in biochemical reactions, in which the catalytic action of an enzyme E binding to a specific

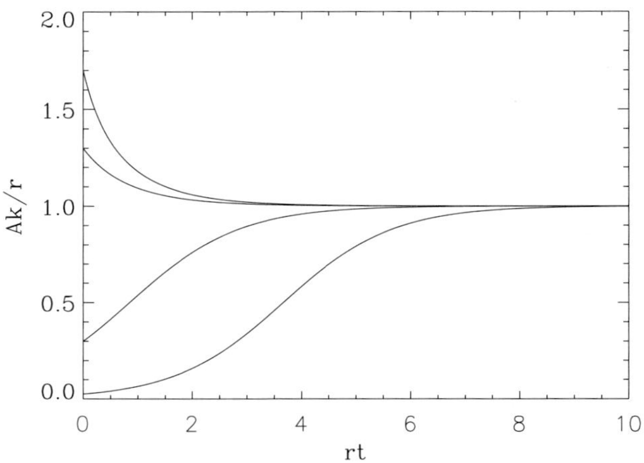

Figure 3.1: Solutions (3.24) of the logistic differential equation for different initial conditions. $r/k = Q_3$ for (3.9) and $r/k = Q$ for (3.23).

substrate S is necessary to transform it into a product P. Michaelis and Menten (1913) modelled the process with this set of elementary chemical reactions:

$$S + E \underset{k_2}{\overset{k_1}{\rightleftharpoons}} C , \quad C \xrightarrow{k_3} P + E . \tag{3.27}$$

It involves the reversible formation of an intermediate complex C, which leads later irreversibly to product formation. The kinetic equations are

$$\frac{dS}{dt} = -k_1 SE + k_2 C \tag{3.28}$$

$$\frac{dE}{dt} = -k_1 SE + (k_2 + k_3)\, C \tag{3.29}$$

$$\frac{dC}{dt} = k_1 SE - (k_2 + k_3)\, C \tag{3.30}$$

$$\frac{dP}{dt} = k_3 C . \tag{3.31}$$

The variable P only appears in (3.31) so that $P(t)$ can be obtained from this last equation once the dynamics of $C(t)$ has been determined from (3.28)-(3.30). In addition, adding Eqs. (3.29) and (3.30) one finds the conservation law

$$E(t) + C(t) = E(0) + C(0) \equiv E_T \qquad (3.32)$$

The notation E_T denotes the "total" enzyme in the system, i.e. the sum of the one in free form and in the complex. From (3.32) we can eliminate E in (3.28) and (3.30) to obtain

$$\frac{dS}{dt} = -k_1 E_T S + (k_1 S + k_2) C \qquad (3.33)$$

$$\frac{dC}{dt} = k_1 E_T S - (k_1 S + k_2 + k_3) C . \qquad (3.34)$$

Now an important observation can be done (Briggs and Haldane, 1925): Enzymes are so efficient that a very small amount is enough to catalyze the reaction. Thus, except at very late times when the substrate is nearly exhausted, we have that E, C, and E_T are small parameters when compared to S, and the second enzyme dynamics equation, (3.34), evolves in a time scale much faster than the first. This situation of separation of time scales allows the use of the *adiabatic approximation*, or *adiabatic elimination*, a useful tool which will be used several times in the following. The method applies to systems of equations of the form

$$\frac{dx}{dt} = \epsilon f(x, y)$$

$$\frac{dy}{dt} = g(x, y) \qquad (3.35)$$

or

$$\frac{dx}{dt} = f(x, y)$$

$$\frac{dy}{dt} = \frac{1}{\epsilon} g(x, y) \qquad (3.36)$$

with ϵ a small parameter, so that the second equation is much faster than the first, except when a neighborhood (of order ϵ) of the null-cline $g(x, y) = 0$ is reached (this occurs within the short time scale of the fast equation, the slow variable remaining nearly constant), after which the two variables become equally slow. On this nullcline, one has a relation $y = y(x)$, which can be substituted into the *slow* equation (3.36) to obtain

$$\frac{dx}{dt} = f(x, y(x)) . \qquad (3.37)$$

If using the form (3.35) instead of (3.36) the r.h.s of (3.37) becomes $\epsilon f(x, y(x))$. Equation (3.37) should be a good approximation to the dynamics after the short time needed for the system to reach the $g(x, y) = 0$ nullcline.

Applying the approximation to the system (3.33)-(3.34), the nullcline of the fast equation ($\dot{C} = 0$) gives

$$C = \frac{k_1 E_T S}{k_1 S + k_2 + k_3} \qquad (3.38)$$

and Eq. (3.33) becomes

$$\frac{dS}{dt} = -k_3 E_T \frac{S}{S + K_M} \quad \left(= -\frac{dP}{dt} \right) \qquad (3.39)$$

where $K_M = (k_2 + k_3)/k_1$ is known as the Michaelis constant. Equation (3.39) contains a nonlinear rate that describes the substrate uptake by the presence of the enzyme, E_T. Even when the substrate is available in large amounts, a saturation in the production rate of product occurs, which has its origin in the limited amount of catalyst. This kinetic result can not be guessed from the stoichiometric expression $S \rightarrow P$. Instead, identification of the elementary steps (3.27) is needed.

In the present context, the adiabatic elimination leading to (3.38) is called the *steady state hypothesis* and was introduced by Briggs and Haldane (1925). In the original treatment by Michaelis and Menten (1913) an assumption of equilibrium for the first reaction in (3.27) was made, which leads to the same result (3.39) but with a different expression for K_M. The steady state hypothesis can be justified rigourously and is precise as long as $E_T \ll K_M + S(0)$ (Segel and Slemrod, 1989).

Many enzymes have more than one binding site, so that they can still catalyze reactions after binding one substrate molecule (an important example is haemoglobin, which has several binding sites for oxygen). A toy model in which the product is produced only after the catalyst binds n substrate molecules (a situation named *cooperativity*) is (See Keener and Sneyd, 1998, Chap. 1):

$$nS + E \underset{k_2}{\overset{k_1}{\rightleftharpoons}} C \ , \quad C \xrightarrow{k_3} P + E \ . \tag{3.40}$$

This leads to rates of the Hill type, which are frequently found experimentally:

$$\frac{dS}{dt} = -K \frac{S^n}{S^n + K_M^n} \quad \left(= -\frac{dP}{dt} \right) \ . \tag{3.41}$$

The important difference with (3.39) is that the rate is very small for small S, being proportional to S^n, and switches quickly to full saturation when $S > K_M$. The larger the n the steepest the switch.

Reaction (3.40) is not very realistic from the molecular point of view. But it provides an approximate description of more plausible mechanisms. For example, catalysis by an enzyme with two binding sites could be modelled by:

$$S + E \underset{k_2}{\overset{k_1}{\rightleftharpoons}} C_1 \ , \quad C_1 \xrightarrow{k_3} E + P$$

$$S + C_1 \underset{k_5}{\overset{k_4}{\rightleftharpoons}} C_2 \ , \quad C_2 \xrightarrow{k_6} C_1 + P \tag{3.42}$$

and this leads (Murray, 1993), when $k_3 \to 0$, to a rate which behaves as the Hill expression Eq. (3.41) with $n = 2$ (which is the behavior experimentally observed in the case of haemoglobin).

As an extreme situation which is related to the rate (3.41) with $n \to \infty$ we mention the *Arrhenius-type* dynamics

$$\frac{dT}{dt} = (1 - T) \exp(-T_c/T) \tag{3.43}$$

in which the rate at small T is smaller than T^n for any value of n. This type of rate appears in simplified models of combustion (Cencini et al., 2003a) in which T is the temperature, an autocatalytic ingredient, but which needs to have a value larger than T_c to start the activated reactions sustaining combustion.

3.1.4 Oscillations and excitability

Oscillatory chemical reactions, once believed to be exotic rarities, or even impossible, are now recognized as being important building blocks in the functioning of any living cell, providing templates for the cell cycle, metabolic activity, or circadian clocks (Goldbeter, 1995). In the inanimate world, an important and versatile group of oscillatory reactions involves the chlorite ion and iodine-containing reactants. Examples are the CIMA (chlorite-iodine-malonic acid) and the CDIMA (chlorine dioxide-iodine-malonic acid) reactions (Epstein and Pojman, 1998). A simplified scheme for the kinetics of these reactions, but capturing the essential features, is

$$
\begin{aligned}
MA + I_2 \;&\rightarrow\; I^- + IMA + H^+ \\
I^- + ClO_2 \;&\rightarrow\; ClO_2^- + \frac{1}{2}I_2 \\
ClO_2^- + 4I^- + 4H^+ \;&\rightarrow\; Cl^- + 2I_2 + 2H_2O \; .
\end{aligned}
\tag{3.44}
$$

MA and IMA are the malonic and the iodomalonic acid, respectively. For the CDIMA case in the usual experimental conditions, it is appropriate to assume that the only dynamical variables are $[I^-]$ and $[ClO_2^-]$, the other concentrations staying essentially constant. In addition, the two first reactions in (3.44) follow the law of mass action, whereas the third one proceeds at a rate which is empirically given by

$$
r = k\frac{[I^-][ClO_2^-]}{a + [I^-]^2} \; .
\tag{3.45}
$$

With these approximations, and after proper rescaling to render the variables dimensionless, an appropriate model for the CDIMA reaction is

$$
\begin{aligned}
\frac{dx}{d\tau} &= \alpha - x - \frac{4xy}{1 + x^2} \\
\frac{dy}{d\tau} &= \beta\left(x - \frac{xy}{1 + x^2}\right) ,
\end{aligned}
\tag{3.46}
$$

where x is proportional to the concentration $[I^-]$ and y to $[ClO_2^-]$. Concentrations oscillate when $\beta < 3\alpha/5 - 25/\alpha$. An example is shown in Fig. 3.2.

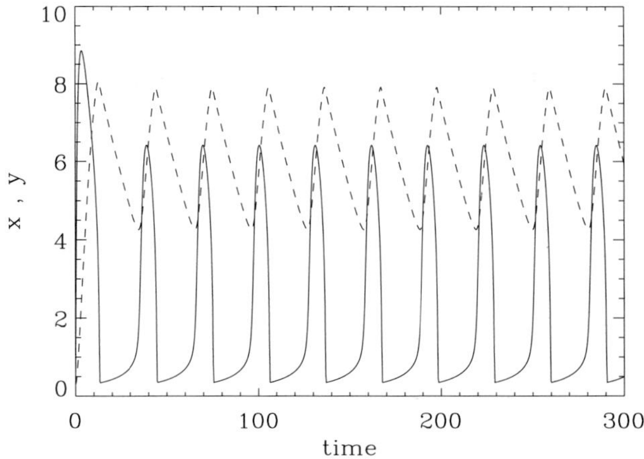

Figure 3.2: Oscillations of the CDIMA model (3.46) for $\alpha = 10$ and $\beta = 0.1$. Solid line is the x concentration, proportional to $[\text{I}^-]$, and dashed is y, proportional to $[\text{ClO}_2^-]$.

Despite the importance of the chlorite-iodide systems in the development of nonlinear chemical dynamics in the 1980s, the Belousov-Zhabotinsky(BZ) reaction remains as the most intensively studied nonlinear chemical system, and one displaying a surprising variety of behavior. Oscillations here were discovered by Belousov (1951) but largely unnoticed until the works of Zhabotinsky (1964). Extensive description of the reaction and its behavior can be found in Tyson (1985), Murray (1993), Scott (1991), or Epstein and Pojman (1998). There are several versions of the reaction, but the most common involves the oxidation of malonic acid by bromate ions BrO_3^- in acid medium and catalyzed by cerium, which during the reaction oscillates between the Ce^{3+} and the Ce^{4+} state. Another possibility is to use as catalyst iron (Fe^{2+} and Fe^{3+}). The essentials of the mechanisms were elucidated by Field et al. (1972), and lead to the three-species model known as the *Oregonator* (Field and Noyes, 1974). In this model the important chemical species are

$$A = \mathrm{BrO_3^-} \ , \qquad P = \mathrm{HOBr} \ , \qquad B = \text{oxidizable organic species}$$

$$X = \mathrm{HBrO_2} \ , \qquad Y = \mathrm{Br^-} \ , \qquad Z = \mathrm{Ce^{4+}} \tag{3.47}$$

and its dynamics is described by the scheme

$$A + Y \xrightarrow{k_1} X + P \ , \quad X + Y \xrightarrow{k_2} 2P \ , \quad A + X \xrightarrow{k_3} 2X + 2Z \ ,$$

$$2X \xrightarrow{k_4} A + P \ , \qquad B + Z \xrightarrow{k_5} \frac{f}{2} Y \ . \tag{3.48}$$

The first two reactions describe the consumption of bromide $\mathrm{Br^-}$, whereas the last three ones model the buildup of $\mathrm{HBrO_2}$ and $\mathrm{Ce^{4+}}$ that finally leads to bromide recovery, and then to a new cycle. By assuming that the bromate concentration $[A]$ remains constant as well as $[B]$, and noting that P enters only as a passive product of the dynamics, the law of mass action leads to

$$
\begin{aligned}
\frac{dX}{dt} &= k_1 AY - k_2 XY + k_3 AX - 2k_4 X^2 \\
\frac{dY}{dt} &= -k_1 AY - k_2 XY + k_5 \frac{f}{2} BZ \\
\frac{dZ}{dt} &= 2k_3 AX - k_5 BZ \ .
\end{aligned}
\tag{3.49}
$$

We follow Tyson (1985) and bring the model to dimensionless form by rescaling time and concentrations:

$$x = \frac{X}{X_0} \ , \quad y = \frac{Y}{Y_0} \ , \quad z = \frac{Z}{Z_0} \ , \quad \tau = \frac{t}{t_0} \tag{3.50}$$

where the scaling factors are:

$$
X_0 = \frac{k_3 A}{2k_4} \qquad Y_0 = \frac{k_3 A}{k_2} \qquad Z_0 = \frac{(k_3 A)^2}{k_4 k_5 B} \qquad t_0 = \frac{1}{k_5 B}
$$

$$
\epsilon_1 = \frac{k_5 B}{k_3 A} \qquad \epsilon_2 = \frac{2k_4 k_5 B}{k_2 k_3 A} \qquad q = \frac{2k_1 k_4}{k_2 k_3} \ . \tag{3.51}
$$

In terms of these variables, the model reads

$$\epsilon_1 \frac{dx}{d\tau} = qy - xy + x(1 - x) \tag{3.52}$$

$$\epsilon_2 \frac{dy}{d\tau} = -qy - xy + fz \tag{3.53}$$

$$\frac{dz}{d\tau} = x - z . \tag{3.54}$$

It can be shown that there is a surface of Hopf bifurcations in the parameter space $(\epsilon_1, \epsilon_2, f)$, so that oscillations in the chemical concentrations are expected at one side of that surface. Nevertheless, analysis is much simplified for interesting experimental situations leading to particular parameter values. To be specific we consider the situation analyzed in Scott (1991), for which $\epsilon_2 \approx 4 \times 10^{-4} \ll \epsilon_1 \approx 0.04$ (q takes values of the order of 10^{-3} and f of order 1). Because of the smallness of ϵ_2, the time derivative in Eq. (3.53) is very large except when the variables sit on the nullcline, $-qy - xy + fz = 0$. This makes y to grow or to decrease fast until the nullcline is reached (more precisely, a neighborhood of size $\mathcal{O}(\epsilon_1)$ of the nullcline). After this moment

$$y = \frac{fz}{q + x} , \tag{3.55}$$

and the other two equations read

$$\epsilon_1 \frac{dx}{d\tau} = F(x, z) \equiv x(1 - x) + \frac{q - x}{q + x} fz \tag{3.56}$$

$$\frac{dz}{d\tau} = G(x, z) \equiv x - z . \tag{3.57}$$

We note that a different reduced set of two equations can be obtained in the alternative case of $\epsilon_1 \ll \epsilon_2$, analyzed for example in (Murray, 1993).

System (3.56)-(3.57) is useful to illustrate a variety of dynamical regimes, from relaxation to a fixed point, to oscillations, and to the occurrence of the excitability phenomenon. Figure 3.3 shows the nullclines $F(x, z) = 0$ and $G(x, z) = 0$. Any steady state should lie on the intersection of those two lines. Again, the presence of a small parameter ($\epsilon_1 \approx 0.04$) multiplying the temporal derivative in

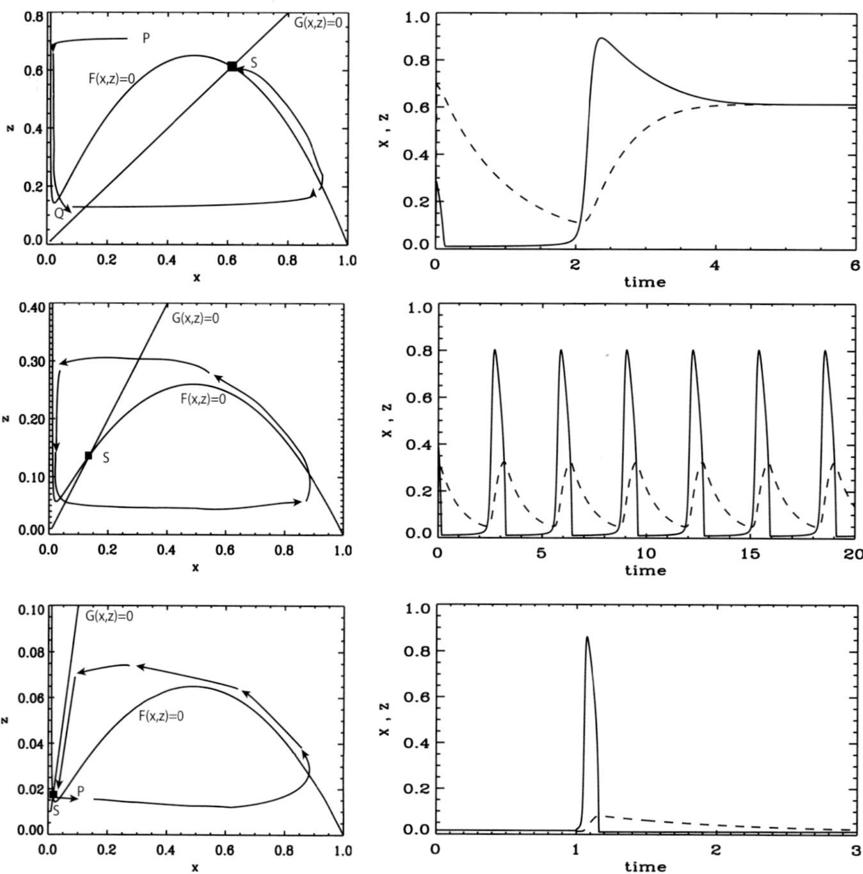

Figure 3.3: Dynamics of the system (3.56)-(3.57). Left panels show the phase space, the two nullclines $F(x, z) = 0$ and $G(x, z) = 0$ and schematic trajectories. The arrows indicate the direction of motion. Right panels show the time evolution of x (solid line) and z (dashed line). In all panels $q = 0.01$, and the square labelled by S indicates the fixed point. First row: $f = 0.4$, $\epsilon_1 = 0.04$. The fixed point is stable and dynamics, starting from P, finally leads to this steady state. Second row: $f = 0.4$, $\epsilon_1 = 0.04$. The fixed point is unstable and concentrations oscillate in a limit cycle. Third row: $f = 4$, $\epsilon_1 = 0.004$. The system lies on the stable fixed point S until time=1. Then its position is slightly displaced in the direction of P (jump not visible in the scale of the right panel), so that a excitation pulse occurs, after which the system returns to the steady state S.

the first equation allows understanding the dynamics of the system in terms of the *adiabatic approximation*. In fact, the value $\epsilon_1 \approx 0.04$ is not small enough for the elimination to provide an accurate approximation to the full dynamics, but the main qualitative features are very well reproduced. In particular, Eq. (3.56) tell us that x grows very fast when the system state is below the nullcline $F(x, z) = 0$, and decreases fast when above. This fast motion persists until $F(x, z) = 0$, or:

$$fz = x(1 - x)\frac{x + q}{x - q} \, , \tag{3.58}$$

is reached. After this, the dynamics will proceed along the nullcline following Eq. (3.57), eventually reaching the fixed point at the intersection with $x = z$, if accessible. Equation (3.57) says that z grows if it is below the line $z = x$, and decreases otherwise. Figure 3.3 presents three qualitatively different situations. In the first row the fixed point S lies to the right of the maximum of the $z = z(x)$ line given by (3.58). For the initial condition P, which lies above the nullcline (3.58), the trajectory starts with a fast jump in the negative x direction until (3.58) is reached. Since the system state is above the other nullcline, $z = x$, the concentrations follow the x-nullcline in the direction of decreasing z. When arriving to the point Q, z should continue decreasing, but the nullcline turns up there. In consequence the trajectory abandons the nullcline, so that there is again a fast jump in x (this time towards the positive direction) until reaching the other branch of the x-nullcline. Since the $z = x$ line has been crossed, the motion on the nullcline is now in the increasing z direction, and the fixed point S is finally reached when $\tau \to \infty$. In the second row the value of f is larger, so that nullcline intersection and the fixed point occurs in a different branch of the x-nullcline. As shown by the plotted trajectories, the dynamics consists of slow decreasing and increasing motions on the lateral branches of the x-nullcline joined by fast jumps in the positive and the negative x direction, respectively. The fixed point is now unreachable (it can also be shown that it is linearly unstable), and a limit cycle leading to periodic concentration oscillations is present. Finally, the third situation (third row) depicted in Fig. 3.3 resembles the one in the first row, in the sense that the dynamics approaches the stable fixed point S at long times. But the difference is that now the fixed point lays very close to the turn-up of the x-nullcline. Thus, a small perturbation applied to the system when it is in S may bring it below the x nullcline, after which a large excursion of the concentrations will occur, before finally approaching and relaxing towards the steady state S. This is called an *excitation* pulse. It is clear that the perturbation could be small, but always larger than a minimum threshold value related to the distance of the fixed point to the $F(x, z) = 0$ line. A system in this situation in which the response to a small (but above threshold) perturbation is a large but transient pulse is said to be *excitable*.

We have seen that the Belousov-Zhabotinsky reaction, even in the restricted parameter range for which some elementary analysis can be done, has a large variety of behaviors, which makes it the ideal model system to illustrate nonlinear dynamics of chemical systems. We briefly mention here a kinetic system of a rather different origin, the FitzHugh-Nagumo (FN) model (Murray, 1993; Meron, 1992):

$$\frac{dx}{dt} = F(x, y) = x(\alpha - x)(x - 1) - y + I \tag{3.59}$$

$$\frac{dy}{dt} = G(x, y) = \epsilon\,(x - \gamma y) \ . \tag{3.60}$$

This model was proposed (FitzHugh, 1961; Nagumo et al., 1962) as a simplification of the Hodgkin-Huxley theory for impulse propagation in nerve membranes (Hodgkin and Huxley, 1952). The variables x and y are related to the membrane potential and to conductances, respectively. At variance with previous chemical models, the dynamical variables here can take negative values. Figure 3.4 shows the shape of the nullclines, $F(x, y) = 0$ and $G(x, y) = 0$. For the parameter values in panels a), b), and c) we see the same qualitative features as in the first rows of Fig. 3.3. Also the small parameter ϵ plays a role similar to ϵ_1 in (3.56)-(3.57): it makes the second equation slow with respect to the first. In consequence, the FN system (3.59)-(3.60) displays qualitative behavior similar to models of the BZ reaction, including oscillations and excitability. Its somewhat simpler algebraic structure makes it a suitable model for more detailed qualitative studies of chemical dynamics, in particular in the context of excitability, even if its original derivation is in a different setting.

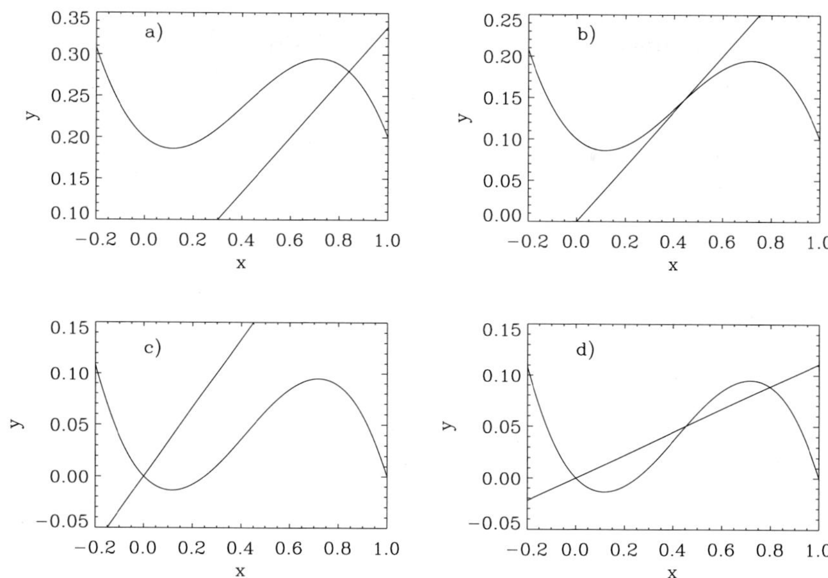

Figure 3.4: Nullclines of the FitzHugh-Nagumo model (3.56)-(3.57). The straight line is $G(x, y) = 0$, and the curve is $F(x, y) = 0$. Fixed points are at their intersections. In all panels $a = 0.25$. a) $\gamma = 3$, $I = 0.2$. There is only one fixed point, which is stable, and the dynamics is qualitatively similar to the one in the first row of Fig. 3.3. b) $\gamma = 3$, $I = 0.1$. There is a single unstable fixed point. For small ϵ, say $\epsilon \lesssim 0.04$, the behavior is oscillatory and qualitatively similar to the one displayed in the second row of Fig. 3.3. c) $\gamma = 3$, $I = 0$. There is a single stable fixed point, which is close to the lower turning point of the $F(x, y) = 0$ nullcline. For small ϵ the dynamics is excitable, as in the third row of Fig. 3.3. d) $\gamma = 9$, $I = 0$. There are three fixed points. The central one is unstable and the other two are linearly stable. Thus the system presents bistability.

3.1.5 Multistability

The last panel in Fig. 3.4 shows a situation qualitatively different from the previous ones. Here, the FN nullclines intersect at three points. Simple linear stability analysis reveals that the central one is unstable, whereas the lateral ones are both stable. Since there are no more attracting structures in this simple dynamical system, the final state of the FN dynamics will be one of these stable fixed points, depending on the initial condition. This situation is called *bistability*.

The simplest chemical scheme leading to this behavior is a hypothetical cubic autocatalysis in a CSTR, i.e. one described by the chemical reaction

$$A + 2B \rightarrow 3B \tag{3.61}$$

at rate k_1 coupled with input and output of the reacting species into the reactor as in (3.15), at rate k_0. The kinetic equations are

$$\begin{aligned}
\frac{dA}{dt} &= k_0(A_0 - A) - k_1 AB^2 \\
\frac{dB}{dt} &= k_0(B_0 - B) + k_1 AB^2 \; .
\end{aligned} \tag{3.62}$$

The equation for $A + B$ is linear and can be readily solved, showing that at long times $A + B$ approaches $A_0 + B_0$. If we start already with this initial condition, $A(0) + B(0) = A_0 + B_0$, then $A(t) + B(t) = A_0 + B_0 \; \forall t$, and $B(t) = A_0 + B_0 - A(t)$. We assume that only A is fed into the system, i.e. $B_0 = 0$. Then, the first equation in (3.62) becomes

$$\begin{aligned}
\frac{dA}{dt} \equiv F(A) &= k_0(A_0 - A) - k_1 A (A_0 - A)^2 \\
&= (A_0 - A)(k_0 - k_1 A(A_0 - A)) \; .
\end{aligned} \tag{3.63}$$

This equation always has a stable fixed point at $A = A_0$, the *unreacted* state. If $k_1 A_0^2 > 4k_0$ there are in addition two new fixed points: $A_\pm = (2k_1)^{-1} \left(k_1 A_0 \pm (k_1^2 A_0^2 - 4k_0 k_1)^{1/2} \right)$. A_+ is linearly stable and A_- linearly unstable. Thus, for $k_1 A_0^2 > 4k_0$ the system is bistable.

Note the similarity with Eq. (3.59) as a function of x. The r.h.s. of both are simple cubic polynomials, which when having real roots can be generically written as $F(x) = \mu(x-x_1)(x-x_2)(x-x_3)$. In fact the FN model (3.59)-(3.60) can be thought to be a nearly-bistable system in which the slow variable y jumps between the upper and the lower x equilibria. This point of view has been fruitful in designing chemical oscillators starting from bistable systems (Epstein and Pojman, 1998). There are many real chemical reactions displaying bistability in adequate parameter ranges(Epstein et al., 1981), among which we cite the chlorite-iodine and the arsenite-iodate systems.

In addition to bistability, the situation in which any of three stable states can be reached depending on the initial conditions – *tristability*– has also been reported in chemical systems (Orban et al., 1982). One can also have multistability among different types of cyclic behavior (Goldbeter, 1995) and other types of dynamics.

We close this Section by mentioning that, despite initial controversies, the most complex type of dynamic behavior, chaos, has been shown to be also present in chemical systems, among which the most studied is again the BZ reaction (Scott, 1991). Chaos has also been observed, for example, in the chlorite-thiosulfate or the bromate-chlorite-iodide reactions, or in the gas-phase reaction between carbon monoxide and oxygen (Epstein and Pojman, 1998).

3.2   Biological models

3.2.1   Simple birth, death and saturation

When dealing with a single population, the standard modelling framework is

$$\frac{dP}{dt} = B - D + M \tag{3.64}$$

where P is the number of individuals and B, D, and M are, respectively, the birth rate (number of births per unit of time), the death rate, and the migration rate. Migration is mediated by transport processes that will be the subject of future chapters, so that it will not be considered here. For the birth term the simplest mechanism is the one at work in microorganisms that reproduce by bipartition. If each of them reproduces at a constant rate, the growth rate of the total population is proportional to the number of individuals, so that $B = bP$ and b is a *per capita* growth rate. Note that this is consistent with the application of the law of mass action to the chemical scheme $P \xrightarrow{b} 2P$ (Eqs. (3.19) and (3.20)). In sexually reproducing populations this scheme can still be considered as an approximation to more realistic interactions. In the same way, if each individual has a probability of dying which is constant in time, the death rate is proportional to the population $D = dP$, consistently with the chemical mechanism $P \xrightarrow{d} \emptyset$ (Eq. (3.11)), and d is a *per capita* death rate. We will omit the qualification 'per capita' when no confusion could arise. The combination of these expressions leads to the simple linear equation

$$\frac{dP}{dt} = rP \qquad (3.65)$$

where $r = b - d$ is the net growth rate (*per capita*), or the difference between birth and death rates. Equation (3.65) leads to the Malthusian exponential population growth as long as $b > d$. Volterra (1926) proposed expressions for B and D in terms of the populations involved by using the law of mass action. The argument is that predation, death, reproduction, etc. are events of similar nature to the molecular collisions that lead to the law of mass action. As an example, a mechanism that can stop the Malthusian explosion is the finiteness of the resources. If we consider that reproduction requires food, a scheme of the type $A + B \rightarrow 2A$ is appropriate, where B represents the food. The kinetics of this process is given by (3.23). In the biological context it is more common to write this as

$$\frac{dP}{dt} = rP \left(1 - \frac{P}{K} \right) \equiv g(P)P \ . \qquad (3.66)$$

Here r, the difference between birth and death rates is the maximum growth rate, since the effective population-dependent growth rate $g(P) = r(1 - P/K)$ decreases with P. K is the carrying capacity, identified with the maximum population that can be sustained by the available resources, that is the population value at which growth stops.

The growth rate of P in (3.66) is maximum when $P \rightarrow 0$ and very small populations always grow. Of course, if the population is so small that there is less than a single individual this is no longer valid, and a continuous and deterministic description is not appropriate. In stochastic models of discrete populations (Renshaw, 1991) typically one finds that small populations have great chances to become extinct, even if the average growth rate is positive. Also, in sexually reproducing populations, a small number of individuals will experience difficulties in finding mate. All these mechanisms, and others arising from predation, which will reduce growth at small populations, are collectively known as the Allee effect (Allee, 1938; Dennis, 1989). A way to model it is to modify the per capita growth rate $g(P)$ in (3.66) with a function that decreases when approaching $P \to 0$, so that the maximum is not at $P = 0$ but at some intermediate (usually small) population value. The simplest such form for the growth rate is the quadratic function: $g(P) = \mu(P - b)(K - P)$ so that the population dynamics is

$$\frac{dP}{dt} = g(P)P = \mu P(P - b)(K - P) \; . \tag{3.67}$$

When $b > 0$, the growth rate at zero population is negative, a situation called *strong* Allee effect. We have already encountered this dynamics in Sect. 3.1.5. It leads to bistability, with two stable fixed points for the dynamics, $P = 0$ and $P = K$, which will be reached depending on the initial population being below or above the *critical density* b. When $b < 0$, we have *weak* Allee effect, in the sense that growth is slow for small populations, but the origin is always unstable and the only stable fixed point is the one in which the population reaches the carrying capacity $P = K$.

3.2.2   Predator-Prey models

Volterra (1926), in order to explain observations of changes in fish catches in the Adriatic Sea during the First World War, proposed a model in which the growth rate of a prey P in (3.65) is decreased proportionally to the presence of a predator Z. On the other hand the growth rate of a predator Z, negative when alone, is increased proportionally to the availability of the prey:

$$\frac{dP}{dt} = (a_1 - a_2 Z)P \tag{3.68}$$

$$\frac{dZ}{dt} = (b_1 P - b_2)Z . \tag{3.69}$$

It turns out that this model is equivalent to the one derived by Lotka (1920) as arising from application of the law of Mass Action to the hypothetical chemical scheme:

$$P \xrightarrow{a_1} 2P , \quad P + Z \xrightarrow{a_2} 2Z , \quad Z \xrightarrow{b_2} \emptyset . \tag{3.70}$$

These processes correspond, respectively, to spontaneous reproduction of the prey, reproduction of the predator mediated by prey consumption, and predator death. In fact one obtains Eqs. (3.68)-(3.69) with $a_2 = a_1$, but this is just a consequence of measuring the amount of chemicals in moles or molecules, and it expresses the fact that in (3.70) the consumption of each molecule of P leads to a new molecule of Z. Changing to other units such as mass makes the coefficients of the nonlinear terms to become different as in (3.68)-(3.69). In biological settings, common units for either P and Z are either mass or number of individuals, and in these units it is no longer true that ingestion of one unit of P produces one unit of new Z. Lotka presented his model as a chemical scheme that would produce persistent chemical oscillations.

The solutions of (3.68)-(3.69) for positive parameters and generic initial conditions are oscillations, of amplitude fixed by the initial conditions, around the fixed point $Z = a_1/a_2$, $P = b_2/b_1$. The equation of this family of closed trajectories is $a_1 \ln Z + b_2 \ln P - b_1 P - a_2 Z = $ constant. The oscillations are suggestive of the population oscillations observed in some real predator-prey systems, but they suffer from an important drawback: the existence of the continuous family of oscillating trajectories is *structurally unstable*: systems similar to (3.68)-(3.69) but with small additional terms either lack the oscillations, or a single limit cycle is selected out of the continuum. Thus, the model (3.68)-(3.69) can not be considered a robust model of biological interactions, which are never known with enough precision to rule out additional terms in (3.68)-(3.69). The Lotka-Volterra model also has important deficiencies from biological point of view: The first one is that in the absence of predators, the prey population grows unbounded. Resource limitations will be relevant and should be included at high prey densities. Second, the predators are insatiable, in the sense that they always predate at a rate proportional to prey density, independently on how large it is. Some limit to the maximum amount of prey the predator can process per unit of time should be present in realistic situations.

A common way to address the first caveat is to replace the growth term $a_1 P$ for the prey in (3.68) by a more realistic logistic growth (3.66). For the second, it is natural to replace the interaction terms ZP in (3.68)-(3.69) by more general forms. This leads to

$$\frac{dP}{dt} = r\left(1 - \frac{P}{K}\right)P - af(P)Z \tag{3.71}$$

$$\frac{dZ}{dt} = bf(P)Z - cZ \ . \tag{3.72}$$

In the Lotka-Volterra model (3.68)-(3.69), $f(P) = P$. Model (3.71)-(3.72) is by no means the only possibility. Another popular choice is to assume that the functional response f depends not only on the prey, but it is ratio-dependent, i.e. it depends on the *amount of prey per predator* $f = f(P/Z)$, modelling competition for food among predators, that is absent in the previous models. An example is:

$$\frac{dZ}{dt} = s\left(1 - g\frac{Z}{P}\right)Z \tag{3.73}$$

which can also be understood as a model in which the prey determines the carrying capacity (P/g) for Z.

Returning to the prey-dependent functional response $f(P)$, three qualitatively different forms were proposed by Holling (1959a). They are plotted in Fig. 3.5. Type I curve can be considered a limit case of type II. This type II, or *cyrtoid* curve, represents a linear dependence of $f(P)$, as in the case of Lotka-Volterra, for small prey density, but a saturation at a maximum value M at larger values of $P > P_0$, indicating predator satiation or incapacity to process more food.

Specific models with this qualitative behavior are the Holling *disk equation* (Holling, 1959b):

$$f(P) = M \frac{P}{P_0 + P} \tag{3.74}$$

or the one proposed by Ivlev (1961):

$$f(P) = M \left(1 - e^{-\frac{P}{P_0}} \right) . \tag{3.75}$$

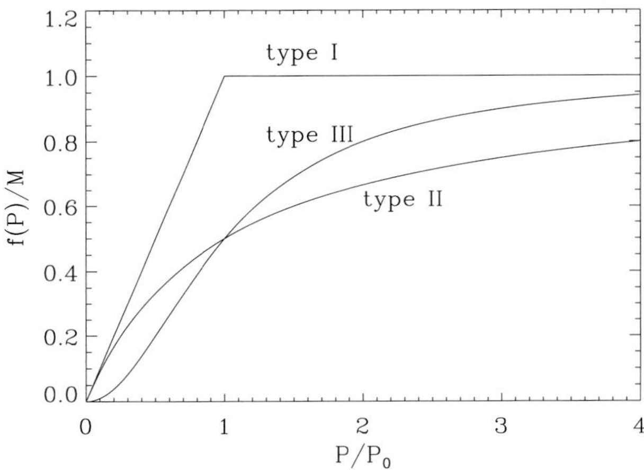

Figure 3.5: The Holling type I, type II (Eq. (3.74)) and type III (Eq. (3.76)) functional responses, giving the amount of predation as a function of the amount of prey P.

The term $f(P)Z$ in (3.71), when $f(P)$ is given by (3.74), leads formally to the Michaelis-Menten dynamics (3.39), if E_T is identified with the predator density and P with the substrate. This analogy has been elaborated in the literature. For example Real (1977) describes predator-prey dynamics with the Michaelis-Menten scheme (3.27), with S the prey, C the intermediate state of the prey when it is eaten, E is the predator searching for food and P is the new predator biomass produced during the consumption process, so that $E_T = E + P$ is the total amount of predator. This leads to a *justification* of the interaction terms in (3.71)-(3.72) by following the steps leading to (3.39). In this context, a predator acts as a specialized enzyme transforming the prey into processed food that is later assimilated. The analogy is not complete, however, since in the Michaelis-Menten dynamics, E_T is constant, whereas (3.71)-(3.72) are intended to be applied to time-dependent Z. Another justification, provided by Holling (1959b), realizes that, in absence of other processes, $P_T \equiv af(P)T$ can be interpreted as the number of preys consumed per predator in a short time T. This quantity should be proportional to the available prey density P but also to the time available to search for food. This time in turn is the total time T minus the time needed to chase, eat, and assimilate each prey, τ, multiplied by the number of consumed preys P_T. Thus $P_T = \kappa P(T - P_T \tau)$ that, after solving for P_T, leads to (3.74) with $M = (a\tau)^{-1}$ and $P_0 = (\kappa\tau)^{-1}$. In any case there is a large amount of empirical evidence justifying (3.74) for microbial and invertebrate predators, dating back to Monod (1942) (Holling, 1959a,b; Waltman, 1983).

Empirical data for vertebrate predators or even higher microbial forms of life display rather a Holling III type sigmoidal shape (Fig. 3.5) that can be modelled with the function

$$f(P) = M\frac{P^2}{P_0^2 + P^2} \, . \tag{3.76}$$

The main differences are the very slow feeding response of the predator ($\propto P^2$) at small prey density, and the sudden increase to saturation for $P > P_0$. This can be attributed to different kinds of *learning* behavior, such as the lack of interest of the predator on a food too scarce and too difficult to obtain, or to the switch to a different food on generalists. Again, the similarity with cooperative enzyme dynamics (3.41) is evident and Real (1977) relates (3.40) and (3.41) with the behavior of a predator that needs n encounters with a prey to reach optimum efficiency at utilizing that prey item as a resource.

We note that all predators except the higher ones are in turn prey for larger species, say F. Thus, terms of the form $-hg(Z)F$, with $g(Z)$ an additional functional response, should be added to (3.72) to represent predation from a higher trophic level and couple it to the equation for F in a larger *food chain*. To close the chain at some level, one has to substitute some *ansatz* for the density of the highest predator, say F, usually in terms of its prey $F = F(Z)$ (this function is usually called the *numerical response* of F to Z, to distinguish it from the *functional responses* f or g). It is typical to use $F = cZ^{r-1}$, and the simple Lotka-Volterra functional response $g(Z) \propto Z$, so that the extra death term in (3.72) becomes $-kZ^r$. Common closure exponents r are $r = 1$ or $r = 2$.

As a model incorporating some of the above ingredients we have the following PZ (for phytoplankton(P)-zooplankton(Z)), with Holling type-III predation of zooplankton on phytoplankton:

$$\frac{dP}{dt} \;=\; F(P, Z) = r\left(1 - \frac{P}{K}\right)P - a\frac{P^2}{P_0^2 + P^2}Z \qquad (3.77)$$

$$\frac{dZ}{dt} \;=\; G(P, Z) = b\frac{P^2}{P_0^2 + P^2}Z - cZ \ . \qquad (3.78)$$

The nullclines of that model are shown in Fig. 3.6 for several parameter values. We see the remarkable similarity with the reduced Oregonator or the FitzHugh-Nagumo models of Sect. 3.1.4 Also, the dynamics of phytoplankton is faster than zooplankton ($r \gg b, c$), so that again the separation in time scales that facilitated analysis in those cases can be used here. As a consequence, we have also in model (3.77)-(3.78) regions in parameter space leading to relaxation to a fixed point, with steady phytoplankton and zooplankton concentrations, regions with predator-prey oscillatory behavior, and also excitable behavior. These features are quite common in models of the general form (3.77)-(3.78) but with different higher-predator closures and variations of the Holling-III predation. If the predation at small prey behaves as the Holling-II type instead, fixed point and oscillatory behavior are also found, but excitability disappears. The relevance of excitability for the understanding of plankton blooms and red tides was first pointed out from model (3.77)-(3.78) by Truscott and Brindley (1994).

We note that, in the limit in which the Z dynamics is so slow that Z can be considered constant, Z_c, there is a range of values of Z_c for which the first equation (3.77) is a bistable dynamical system, with coexistence of a low-P and a large-P steady states (May, 1977).

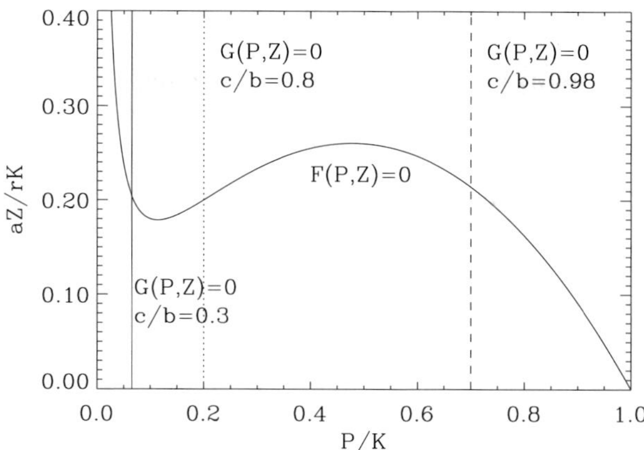

Figure 3.6: Nullclines of the PZ model (3.77)-(3.78). Phytoplankton (P) is shown in units of the carrying capacity K and zooplankton in units of rK/a. $P_0/K = 0.1$. The curve is the nullcline of (3.77), $F(P, Z) = 0$. The vertical lines are the nullclines of (3.78), $G(P, Z) = 0$, for $c/b = 0.98$ (dashed line, dynamics leads asymptotically to a coexistence fixed point), $c/b = 0.98$ (dotted line, the fixed point at nullcline intersection is unstable, and dynamics consists on oscillations around it), and $c/b = 0.3$ (solid line, there is a stable fixed point which is excitable if $r >> b$). Note the correspondence with the three first panels in Fig. 3.4, and with Fig. 3.3.

3.2.3   Competition

Competition among two species means that the increase in one of the populations decreases the net growth rate of the second one, and vice versa. This happens when they feed on the same resources, or if they produce substances (toxins) that are toxic for the other species. A classical competition model was also introduced in Volterra (1926), and considered in a more general parameter range by Lotka (1932). It is known as the competitive Lotka-Volterra system:

$$\frac{dB_1}{dt} = B_1 (r_1 - a_{11} B_1 - a_{12} B_2)$$

$$\frac{dB_2}{dt} = B_2 (r_2 - a_{21} B_1 - a_{22} B_2) \tag{3.79}$$

where B_1 and B_2 are the densities of the two competing organisms. We will take $r_1, r_2 > 0$, so that each species reaches a non-vanishing stable equilibrium (at r_1/a_{11} and at r_2/a_{22}) in the absence of the other. a_{11} and a_{22} are the coefficients of intraspecific competition. The presence of one species decreases the growth of the other if $a_{12} > 0$ and $a_{21} > 0$, the case to be considered here. Symbiosis and mutualism can also be represented by taking negative interspecific competition coefficients.

There are always three steady equilibria:

$$(B_1, B_2) = (0,0), (r_1/a_{11}, 0) \text{ and } (0, r_2/a_{22}),$$

and if either

$$a_{12}/a_{22} < r_1/r_2 < a_{11}/a_{21} \tag{3.80}$$

or

$$a_{12}/a_{22} > r_1/r_2 > a_{11}/a_{21} \tag{3.81}$$

then there is a fourth equilibrium point in the positive quadrant:

$$(B_1^*, B_2^*) = \left(\frac{r_1 a_{22} - r_2 a_{12}}{|A|}, \frac{r_2 a_{11} - r_1 a_{21}}{|A|} \right) \tag{3.82}$$

where $|A| \equiv a_{11}a_{22} - a_{12}a_{21}$ is positive in case (3.80) and negative in case (3.81). For positive r_1 and r_2, the origin is unstable. In case (3.80) the situation is depicted in Fig. 3.7a): The two fixed points at the coordinate axis, representing extinction of one of the two species, are unstable, and the coexistence of the two species (3.82) is the stable attractor of the dynamics. Note that (3.80) implies $a_{11}a_{22} > a_{12}a_{21}$, that can be interpreted as a stronger intraspecific than interspecific competition, and the reverse is true for (3.81). In this last situation the phase portrait is depicted in Fig. 3.7b): (B_1^*, B_2^*) is now a saddle point and at long times one species or the other disappears, depending on the initial conditions. We have a *bistable* situation, in which two possible outcomes are possible at long times. When neither (3.80) nor (3.81) is satisfied, the nullclines do not intersect in the positive quadrant, so that no coexistence fixed point exists. There are only fixed points on the coordinate axes, only one of them being stable, so that at long times only one of the species survives.

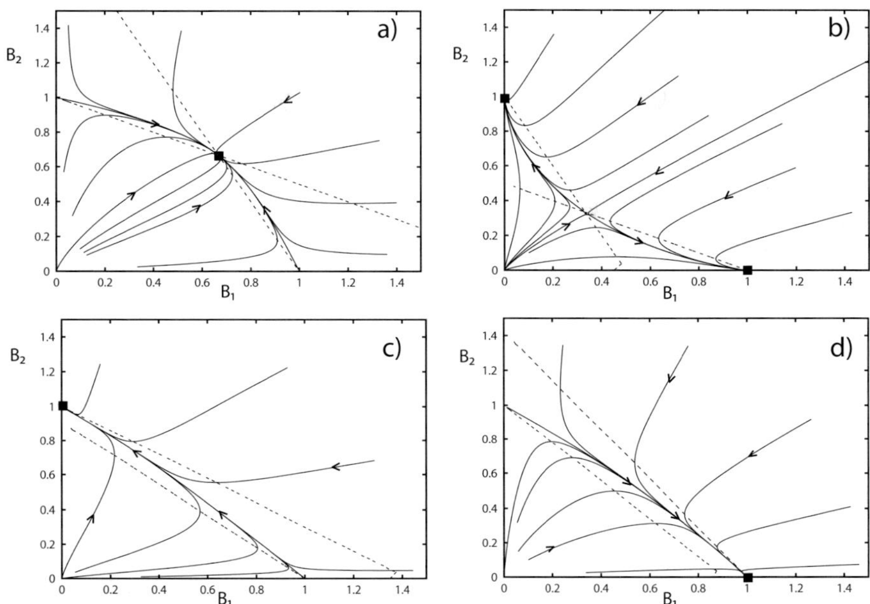

Figure 3.7: Phase portraits for system (3.79). In all panels $a_{11} = a_{22} = 1$, $r_1 = 1$ and $r_2 = 0.8$. The black square indicates the stable fixed points. Dashed lines are the nullclines of the two equations (the coordinate axes are also nullclines). A number of trajectories is plotted, with the arrows indicating the direction of motion. a) $a_{12} = a_{21} = 0.5$ (interspecies competition smaller than the intraspecific one) so that (3.80) is satisfied and species coexistence is the asymptotic stable state. b) $a_{12} = a_{21} = 2$ (interspecies competition larger than the intraspecific one), so that (3.81) is satisfied. The interior fixed point is unstable and there is bistability: depending on the initial condition either extinction of B_1 or of B_2 occurs. c) $(a_{12} = 1.1, a_{21} = 0.7)$ and d) $(a_{12} = 0.7, a_{21} = 1.1)$ do not satisfy (3.80)-(3.81), and the nullclines do not intersect in the interior of the positive quadrant. Then the only fixed points are on the axes, and only one is stable, representing extinction of one of the two species.

A different type of competition interaction is the case in which the growth rates of two competitors depend linearly on the same resource $R(B_1, B_2)$ (exploitative competition):

$$\begin{aligned}
\frac{dB_1}{dt} &= B_1\left(a_1 R(B_1, B_2) - d_1\right) \\
\frac{dB_2}{dt} &= B_2\left(a_2 R(B_1, B_2) - d_2\right) .
\end{aligned} \tag{3.83}$$

We assume that the growth rates between brackets in (3.83) are positive for small B_1, B_2, so that extinction of both species will not occur, and that they become negative at large B_1 and B_2, so that the trajectories are bounded. An example of this situation is provided by the chemical scheme

$$\begin{aligned}
A + B_1 \xrightarrow{a_1} 2B_1 \quad &, \qquad B_1 \xrightarrow{d_1} A \\
A + B_2 \xrightarrow{a_2} 2B_2 \quad &, \qquad B_2 \xrightarrow{d_2} A .
\end{aligned} \tag{3.84}$$

From the rate equations associated to (3.84) the quantity $T = A + B_1 + B_2$ is constant, so that the resource $A = T - B_1 - B_2 \equiv R(B_1, B_2)$ satisfies the hypothesis and the dynamics is of the form (3.83) with R linear in B_1, B_2. Since they are a particular case of (3.79) where $r_1 = a_1 T - d_1$, $r_2 = a_2 T - d_2$, $a_{11} = a_{12} = a_1$, $a_{21} = a_{22} = a_2$, and since (3.80) does not hold generally, we know that there will be no coexistence except in the non-generic case in which $d_1/a_1 = d_2/a_2$. One of the competitors will disappear. We can show that the same *competitive exclusion* (Hardin, 1960; Hutchinson, 1961) behavior holds also in the more general case (3.83) with arbitrary $R(B_1, B_2)$. To this end, let us choose constants c_1, c_2 such that $c_1 a_1 + c_2 a_2 = 0$ and define $\alpha \equiv c_1 d_1 + c_2 d_2$, that can be made positive by adequate choice of c_1 and c_2. Eq. (3.83) implies that

$$c_1 \frac{d \ln B_1}{dt} + c_2 \frac{d \ln B_1}{dt} = -\alpha \tag{3.85}$$

and integrating:

$$B_1(t)^{c_1} B_2(t)^{c_2} = C e^{-\alpha t} \tag{3.86}$$

which vanishes as $t \to \infty$. Since all the trajectories are bounded, this implies that at least one of the densities vanishes at long time. Thus coexistence is impossible.

Following analogous arguments, it can be shown that if there are more species than resources ($s > r$) in a set of equations of the form

$$\frac{dB_i}{dt} = B_i \left(a_{i1} R_1 + \cdots + a_{ir} R_r - d_i \right) \, , \quad i = 1, \ldots, s \qquad (3.87)$$

(note the linear dependence of the growth rate on the resources) then competitive exclusion also occurs. The constants $c_1, \ldots, c_s$ are chosen as solutions of $\sum_{i=1}^{s} c_i a_{ij}$, with $j = 1, \ldots, r$ which can be solved generically when $s > r$, and $\alpha = \sum_{i=1}^{s} c_i d_i$. One obtains $\prod_{i=1}^{s} B_i(t)^{c_i} = C \exp(-\alpha t)$, which again implies that at least one of the populations disappears at long times.

When the dependence of the s growth rates on the $r < s$ resources is not simply linear as in Eq. (3.87), but more general:

$$\begin{aligned} \frac{dB_i}{dt} &= B_i f_i \left(R_1, \ldots, R_r \right) \, , \quad i = 1, \ldots, s \\ R_k &= R_k \left(B_1, \ldots, B_s \right) \, , \quad k = 1, \ldots, r \end{aligned} \qquad (3.88)$$

competitive exclusion does not hold in general (McGehee and Armstrong, 1977), except in the case $s = 2$ considered above (Eq. (3.83)). But, generically, coexistence can not occur simply as fixed points of (3.88) representing steady species coexistence: the coexisting species should oscillate or display more complex dynamic behavior (examples including chaotic behavior can be found in Huisman and Weissing (1999)). This is so because solving for fixed points of (3.88) involves s equations for $r < s$ unknowns, which in general has no solution. This is a form of competitive exclusion which is weaker than the one applying to models (3.83) or (3.87), since it only forbids steady coexistence, but more general since it applies to all equations of the form (3.88). As stressed by Lotka (1932), steady coexistence is possible in the more general model (3.79) if condition (3.80) is satisfied.

Additional examples of competitive behavior can be found in Waltman (1983). The range of possible behavior greatly expands when considering more than two species (Hofbauer and Sigmund, 1988; Huisman and Weissing, 1999; Pigolotti et al., 2007). We just present here another type of competition called *cyclic competition* (Hofbauer and Sigmund, 1988). The classical example is provided by the May-Leonard system (May and Leonard, 1975) for three competing species B_1, B_2, and B_3:

$$
\begin{aligned}
\dot{B}_1 &= B_1\left(1 - B_1 - aB_2 - bB_3\right) \\
\dot{B}_2 &= B_2\left(1 - B_2 - aB_3 - bB_1\right) \\
\dot{B}_3 &= B_3\left(1 - B_3 - aB_1 - bB_2\right)
\end{aligned}
\tag{3.89}
$$

with $0 < b < 1 < a$ and $a + b > 2$. When one of the species is absent, say 3, the dynamics for the other two is familiar from (3.79): species 2 wins and 1 disappears. But the structure of (3.89) implies that in a competition between 2 and 3, 3 is the survivor, and between 3 and 1, 1 wins. Thus, as in the classical rock-paper-scissors game, there is no clear outstanding competitor, and it is not obvious what will happen when the three species are present together. Although by no means in such a symmetric form as (3.89), this situation happens in some real ecosystems, as for example the competition between three different strains of the microorganism E. coli (Kerr et al., 2002) or in the mating strategies of side-blotched lizards (Sinervo and Lively, 1996).

The system (3.89) has five fixed points in the positive octant: The origin $(B_1, B_2, B_3) = (0, 0, 0)$ which is always unstable. The point (z, z, z), with $z = (1 + a + b)^{-1}$, which is a saddle-focus point, with its one-dimensional stable manifold approaching the point along the main diagonal $B_1 = B_2 = B_3$ and unstable spiral orbits escaping from it. The other three fixed points are $(1, 0, 0)$, $(0, 1, 0)$, and $(0, 0, 1)$. Each one represents survival of just one of the species and are unstable saddle points, with two stable directions approaching each point and one unstable, pointing towards the next single species point in the cyclic order $1 \to 2 \to 3 \to 1$. This set of manifolds thus forms a heteroclinic cycle (see Fig. 3.8). At long times typical trajectories approach the heteroclinic cycle and spend increasing amounts of time close to each of the single-species fixed points, with apparent transient death of two of the species for long times, followed by repeated revivals, leading to a perpetual succession of the three species in cyclic order (see Fig. 3.8).

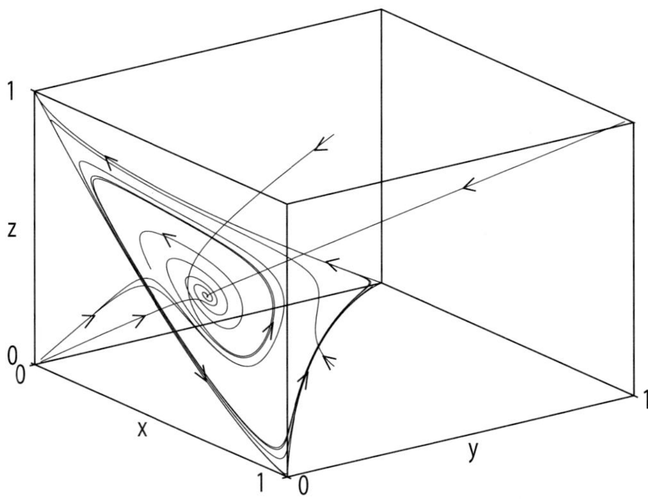

Figure 3.8: The phase space of the May-Leonard system (3.89) for $a = 2$, $b = 0.5$. Some trajectories are shown, with the direction of motion indicated by the arrows. At long times, the trajectories visit the fixed points $(1, 0, 0)$, $(0, 1, 0)$, and $(0, 0, 1)$ in cycling order and spend increasing times in their proximity.

3.3 Summary

We have seen that chemical and biological interactions lead to mathematical models displaying a variety of linear and nonlinear behavior: relaxation to fixed points, multistability, excitability, oscillations, chaos, etc. Despite the different origin of the models, and the diverse nature of the variables they represent (chemical concentrations, population numbers, or even membrane electric potentials) the mathematical structures are quite similar, and it is possible to understand some aspects of the dynamics in one field (e.g. the chemical oscillations in the BZ reaction) with the help of models from other fields (for example the FN model of neurophysiology, or a phytoplankton-zooplankton model). This possibility of common mathematical description will be used in the rest of the book to highlight the similarities and relationships between chemical and biological dynamics when occurring in fluid flows.

Chapter 4

Reaction-diffusion Dynamics

We now consider the effect of diffusive transport in some of the chemical and biological dynamics discussed in the previous Chapter. The phenomenology becomes greatly expanded. In this Chapter we summarize some of the most generic new phenomena.

4.1   Diffusion and linear growth

The simplest combination of reaction and diffusion is the case in which the reaction is a simple linear growth as in (3.20) or (3.65):

$$\frac{\partial C}{\partial t} = \mu C + D\nabla^2 C \ .$$ (4.1)

The exact solution of (4.1) in an unbounded domain is

$$C(\mathbf{x}, t) = \int d\mathbf{x}' G_\mu(\mathbf{x} - \mathbf{x}', t) C_0(\mathbf{x}') \ .$$ (4.2)

The propagator G_μ reads

$$G_\mu(\mathbf{x}, t) = \frac{e^{\mu t} e^{\frac{-\mathbf{x}^2}{4Dt}}}{(4\pi Dt)^{d/2}} \ ,$$ (4.3)

which shows that in this case growth and diffusion are completely uncoupled. At long times, the population explodes everywhere if $\mu > 0$, and approaches extinction if $\mu < 0$. Despite the superficial similarity between the corresponding formulas, the spatial propagation of localized perturbations described by (4.3) when $\mu > 0$ is quite different from the diffusive behavior of (2.11), as we show in the following.

4.1.1 Linear spreading of perturbations

Let's consider first the one-dimensional case (Ebert and van Saarloos, 2000; van Saarloos, 2003):

$$\frac{\partial C}{\partial t} = \mu C + D\frac{\partial^2 C}{\partial x^2} \ . \tag{4.4}$$

If the initial condition is sufficiently localized around the origin then, after a transient (the time needed for (4.3) to become substantially broader than the extent of the initial condition), the full solution (4.2) is well approximated by (4.3). Thus, in this one-dimensional case with localized initial condition we have

$$C(x,t) \propto G_\mu(x,t) = \frac{e^{\mu t} e^{\frac{-x^2}{4Dt}}}{(4\pi Dt)^{1/2}} \ . \tag{4.5}$$

This function is plotted in Fig. 4.1. If $t \to \infty$ at fixed x, we see that $C \to \infty$ but if $x \to \pm\infty$ at fixed t we have $C \to 0$. It is clear that at each instant we have a pair of fronts moving in opposite directions and separating the region in which the concentration is still negligible from the one in which it is large and diverging. From now on we consider only the front travelling towards the right. Its position $R(t)$ can be defined from the location of a particular value, say g, of the concentration: $C(R(t),t) = g$. From (4.5) one finds at long times:

$$R(t) = 2t\sqrt{D\mu}\left(1 - \frac{\ln t}{4\mu t} + \mathcal{O}\left(\frac{\ln g}{t}\right)\right) \ . \tag{4.6}$$

Or, defining the front velocity as $v_f \equiv \dot{R}(t)$:

$$v_f = 2\sqrt{D\mu} - \frac{1}{2t}\sqrt{\frac{D}{\mu}} + \mathcal{O}\left(\frac{\ln g}{t^2}\right) \ . \tag{4.7}$$

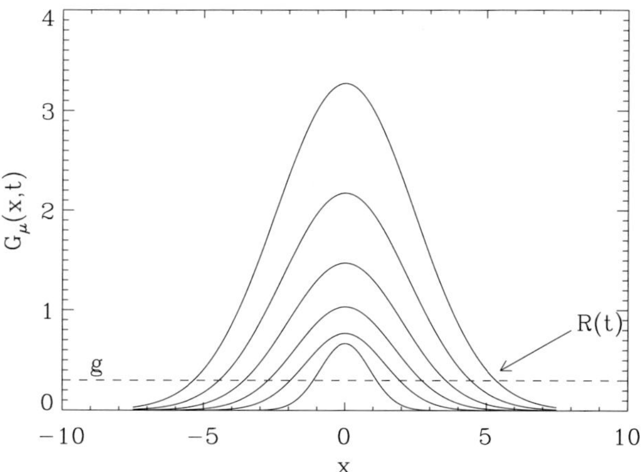

Figure 4.1: The propagator $G_\mu(x, t)$, Eq. (4.5), for $\mu = D = 1$. From bottom to top curve: $t = 0.4, 1, 1.5, 2, 2.5$ and 3. The front position $R(t)$ at the last time and the level g (dashed line) used to define it are also shown.

The dominant terms do not depend on the level g used to define the front. Thus, associated to a uniform state which is unstable with growth rate μ, there is a front controlling the spreading of the concentration that approaches at long times the asymptotic velocity

$$v_f \xrightarrow{t \to \infty} 2\sqrt{D\mu} \equiv c \ . \tag{4.8}$$

This asymptotic speed is reached *from below* and after a rather slow algebraic transient. To focus on the front region Eq. (4.5) can be written in terms of $\xi \equiv x - 2ct$:

$$C(x, t) \propto e^{-\sqrt{\frac{\mu}{D}}\xi} \frac{e^{\frac{-\xi^2}{4Dt}}}{(4\pi Dt)^{1/2}} \equiv P(\xi, t) \ . \tag{4.9}$$

This function has different parts. In the region close to the moving leading edge of the front (defined as $\xi \ll 4Dt$) the solution approaches an exponential of the form $e^{-a_c\xi}$, with $a_c = \sqrt{\mu/D}$. Careful analysis of (4.2) (Ebert and van Saarloos, 2000; van Saarloos, 2003) shows that the condition of "sufficiently localized initial condition" indeed guarantees the validity of (4.9) in the leading edge region, and that precisely means that $C_0(x)$ should decay faster than $e^{-a_c x}$ as $x \to \infty$. In a heuristic way this can be seen from the fact that if $C_0(x)$ decays slower than $e^{-a_c x}$ at infinity, then (4.3) never becomes *broad enough* to justify the approximation of (4.2) by (4.5). The spreading of these *not sufficiently localized* or *too flat* initial conditions is determined by the exponential leading edge and can be found by substituting the ansatz

$$C_a(x,t) = e^{-a(x-v(a)t)} \qquad (4.10)$$

in (4.4). The parameter a is called the steepness of the function. We find

$$v(a) = Da + \frac{\mu}{a}, \qquad (4.11)$$

a velocity always faster than c since the minimum positive velocity is precisely $v(a_c) = c$. Perturbations with a slower than exponential decay have a diverging propagation velocity, and no propagation velocity can be assigned to perturbations which do not decay, i.e. the ones present in the whole system. The functions in (4.10) are all exact solutions of (4.4), but only the ones with $a < a_c$ are relevant to the propagation of generic perturbations with exponential decay, since the steepest ones $(a > a_c)$ are unstable to invasion by the broader (4.3).

The one-dimensional arguments discussed so far are also relevant in higher dimensions. For example, from expression (4.3) one easily shows that in any dimension the front position $R(t) = |\mathbf{x}_f(t)|$ defined from $G_\mu(\mathbf{x}_f(t),t) = g$ always approaches the same asymptotic velocity c, being only slowed in the t^{-1} transient where the factor $1/2$ in (4.7) is replaced by $d/2$.

The propagation of perturbations under growth and diffusion, described by (4.6) or (4.7) is much faster at long times than the diffusive behavior (2.13). Thus, if present, it will always dominate. Of course, an unlimited exponential growth is not possible in real systems and nonlinearities should be added to (4.1). In Sect. 4.2 we will see that the linear results remain relevant in this case. We stress that models similar to (4.1) are not only pertinent to the growth above the zero or empty state, but that they will arise (perhaps in the form of sets of equations, or with higher order derivatives) in linearizations around any homogeneous unstable steady state, that can then be studied with the general tools used here.

4.1.2   The minimum habitat-size problem

Another interesting problem in which (4.4) enters is the situation considered by Skellam (1951), and by Kierstead and Slobodkin (1953) that leads to the so-called KiSS theory. These authors considered the free growth (at rate μ) of phytoplankton (with concentration $C(\mathbf{x}, t)$), in a finite region of water *suitable for growth*, with plankton escaping by diffusion from the suitable zone into poor waters where it immediately dies, so that $C = 0$. The interesting result is that if the suitable region is too small, the growth can not overcome the diffusive escape flux and the patch dies. More generally, this problem will be pertinent to estimate the *minimum habitat size* in the presence of dispersion, modelled as diffusion, and linear growth. The existence of a minimum size can be seen, for example, via the exact one-dimensional solution of (4.1) with boundary conditions $C(x = 0, t) = C(x = L, t) = 0$:

$$C(x, t) = \sum_{n=1}^{\infty} A_n \sin\left(\frac{n\pi x}{L}\right) e^{t\left(\mu - D\left(\frac{n\pi}{L}\right)^2\right)} \quad . \tag{4.12}$$

The constants A_n are fixed by the initial conditions. It is clear that if $L < L_c = \pi\sqrt{D/\mu}$ no growth occurs and the patch dies, so that L_c can be identified with the minimum patch-size able to support growth. Aside from unimportant geometric factors, the same behavior is found in higher dimensions.

There are a number of criticisms to this approach. First, the model is incomplete, since once growth begins it continues without limit. Nonlinear saturation and interactions with predators would be needed to stop this. The diffusion coefficient D certainly does not originate from the Brownian motion of the organisms, since this would be irrelevant to these processes above, say, on the millimeter scale. It is rather a turbulent eddy-diffusion coefficient aimed to represent in an averaged way the effect of dispersion by the turbulent flow, with values that would depend on the observation scale as found for example in Okubo (1971) (See Fig. 2.22). One should ask, therefore, how can the plankton disperse in such a way, whereas the nutrients remain static within the fixed parcel, despite being in the same flow. In any case, if one introduces a growth rate for phytoplankton of the order of one day ($\mu \approx 10^{-5}$ s^{-1}) and an eddy diffusivity appropriate for the 10-100 km range ($D \approx 100$ m^2/s) (Okubo, 1971), one finds $L_c \approx 10$ km, which seemed to compare well with the earliest observations of plankton patches.

4.1.3 Plankton filaments

In the context of plankton dynamics an interesting development that has some common ideas with the KiSS approach but introduces a new and important element is described in Martin (2000). The idea is that dispersion of a patch is not only controlled by eddy diffusivity, but also by the geometric characteristics of the mean flow. It turns out that if an incompressible fluid flow induces dispersion in one direction it necessarily produces convergence in another, to conserve the fluid volume. This was already exploited in Sect. 2.7.1, and includes the ingredient of advection, in addition to the reaction-diffusion processes which are the subject of this Chapter. Nevertheless, since this case can be analyzed easily and extends the KiSS model, we consider it here.

Martin (2000) proposes the following model to describe the transverse profile of a phytoplankton filament:

$$\frac{\partial P}{\partial t} - \lambda x \frac{\partial P}{\partial x} = \mu(t)P + D\frac{\partial^2 P}{\partial x^2} \ . \tag{4.13}$$

This is exactly the filament model (2.87) except for the linear growth term. There are two differences with respect to the KiSS model. One is the possible time-dependence of the growth rate $\mu(t)$, which is a simplified linear representation of nonlinear interactions and predation on phytoplankton following the initial stage of growth. The other is the advective term $-\lambda x \partial_x P$ that models a local strain.

Equation (4.13) can be solved by reducing it to (2.87). It is enough to change $P(x,t) = e^{\int_0^t \mu(u)du}C(x,t)$. The basic globally stable solution (2.89) becomes:

$$P(x,t) = P_0 e^{-\frac{x^2\lambda}{2D}} e^{\int_0^t \mu(u)du - \lambda t} \ . \tag{4.14}$$

It is again a Gaussian with time-dependent height but characteristic width $l_D = \sqrt{D/\lambda}$. At variance with the KiSS approach, the lateral scale of the filament is not controlled by some externally imposed size of the suitable water, but by the competition between diffusion and advection. It is somewhat surprising that the biological growth rate does not affect the filament width in this model. As we will see in Chapter 7 this feature is not shared by more realistic models.

The general solution in the unbounded line can also be found from (2.91):

$$P(x,t) = e^{\int_0^t \mu(u)du - \lambda t} e^{-\frac{x^2\lambda}{2D}} \sum_{n=0}^{\infty} P_n e^{-n\lambda t} H_n\left(x\sqrt{\lambda/2D}\right) \ . \tag{4.15}$$

Again, this solution shows explicitly that the basic filament solution is the dominant structure approached at long times.

4.2   Fisher waves: the invasion of an unstable state by a stable one

Fisher (1937), proposed an equation combining the logistic growth mechanism of (3.9), (3.23) or (3.66) with diffusion to model the spatial spreading of favorable genes in a population distributed in one dimension:

$$\frac{\partial C}{\partial t} = \mu C \left(1 - \frac{C}{K}\right) + D\frac{\partial^2 C}{\partial x^2} \ . \tag{4.16}$$

Careful study of this and of the more general type of equations

$$\frac{\partial C}{\partial t} = F(C) + D\frac{\partial^2 C}{\partial x^2} \tag{4.17}$$

with $F(C)$ satisfying specific conditions was performed in the classical paper by Kolmogorov, Petrovskii, and Piskunov (1937). The conditions are essentially that $F(C)$ is a continuous function having only two zeros (say at C_u and C_s, with $C_u < C_s$) and being positive in between, with $F'(C_u) > 0$ being the largest slope in the interval ($F'(C_u) > F'(C)$, $\forall C \in [C_u, C_s]$). In addition to the parabolic function in (4.16), $F(C) = \mu C(1 - C/K)$, another function of this class used in the original work by Kolmogorov et al. (1937) is $F(C) = \mu C(1 - C)^2$. Figure 4.2 sketches two functions $F(C)$, the parabola in (4.16), and a different one implementing a weak Allee effect (see Sect. 3.2.1) which does not satisfy the above condition on the slope. Equations of the form of (4.16), or (4.17) satisfying the Kolmogorov, Petrovskii, and Piskunov (1937) conditions, are now known as the Fisher equation, the Fisher-Kolmogorov equation, or the FKPP equation. Other important steps in the understanding of Eq. (4.17) and its generalizations were presented by Aronson and Weinberger (1978).

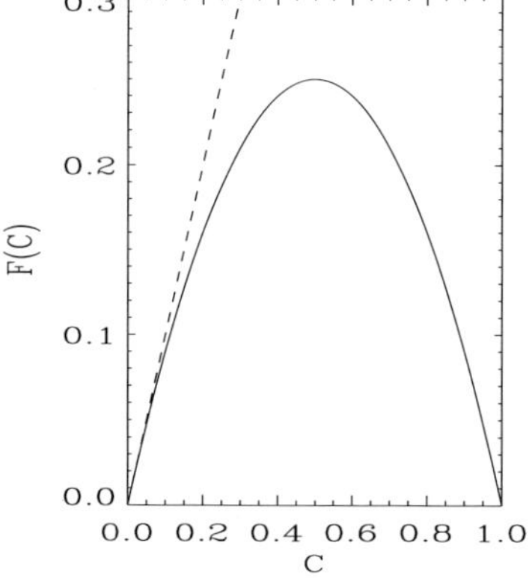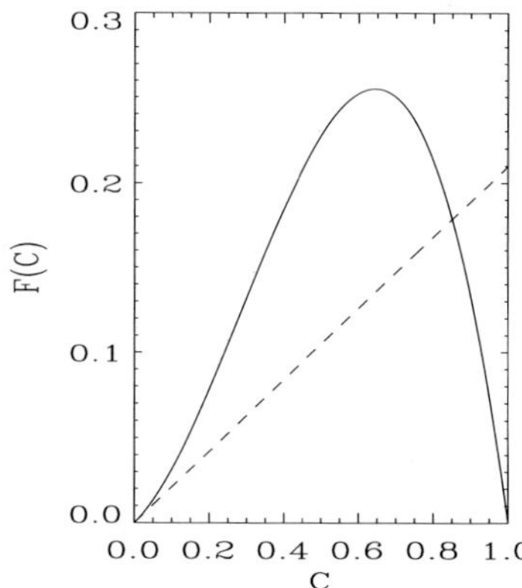

Figure 4.2: Two growth functions $F(C)$ to be used in Eq. (4.17)). The one in the left is the parabolic one in the logistic (4.16), with $\mu = 1$. The right one is of the form (3.67) with $\mu = 1$ and $b = -0.15$, implementing a weak Allee effect. For the first one, the slope at the origin (indicated by the dashed line) is the maximum slope, which is not the case for the function on the right.

The model (4.16) has two spatially uniform steady solutions: $C(x) = 0 \equiv C_u, \forall x$ and $C(x) = K \equiv C_s, \forall x$. The second one is linearly stable and it can be shown that under the usual boundary conditions (e.g. periodic or no flux) it is the only global attractor of the system. C_u is linearly unstable, so that perturbations on top of it will grow, bringing the system eventually to the state C_s. Figure 4.3a shows this process when the initial perturbation is distributed over the whole system. In Fig. 4.3b, however, the evolution is shown when the initial perturbation is sufficiently localized: After an initial transient, two sharp fronts are formed and move in opposite directions at constant speed. They advance into the unstable state $C_u = 0$ replacing it by the stable $C_s = 1$. The process resembles the one depicted in Fig. 4.1 for the linear growth process, but here the growth is saturated by the nonlinear term, and in the region in which each front can be considered as isolated, the solution has the form

$$C(x,t) = g(x - vt) \equiv g(\xi) \ . \tag{4.18}$$

A first step to understand the nature of these fronts is to consider (4.16) on the infinite line and particularize it to the class of solutions (4.18). One obtains a second order ordinary differential equation for $g(\xi)$:

$$Dg'' + vg' + \mu g \left(1 - g/K\right) = 0 \ . \tag{4.19}$$

In order to represent a front between the two states, the pertinent boundary conditions are (for $v > 0$, i.e. a front moving to the right) $g(\xi = -\infty) = C_s$, and $g(\xi = \infty) = C_u$ (which implies $g'(\xi = -\infty) = g'(\xi = \infty) = 0$). This is a nonlinear eigenvalue problem in which one should determine simultaneously the speed v and the function $g(\xi)$. The general method to establish the multiplicity of the solutions involves the consideration of the phase plane (g, h) of the dynamical system of first order differential equations for $g = g(\xi)$, and $h = g'(\xi)$ and its linearization around the fixed points $P_- = (C_s, 0)$ and $P_+ = (C_u, 0)$. This identifies the dimensionality of their stable and unstable manifolds ($d_s^{\pm}$ and $d_u^{\pm}$, where the upper index refers to the points P_+ and P_-). A front solution is a trajectory leaving the point P_- along its unstable manifold in the far past ($\xi = -\infty$) and reaching the point P_+ along its stable manifold in the far future ($\xi = \infty$). The dimension of the intersection of these manifolds ($d_s^- + d_u^+ - d$, where $d = 2$ is the phase space dimension) determines the dimension of the

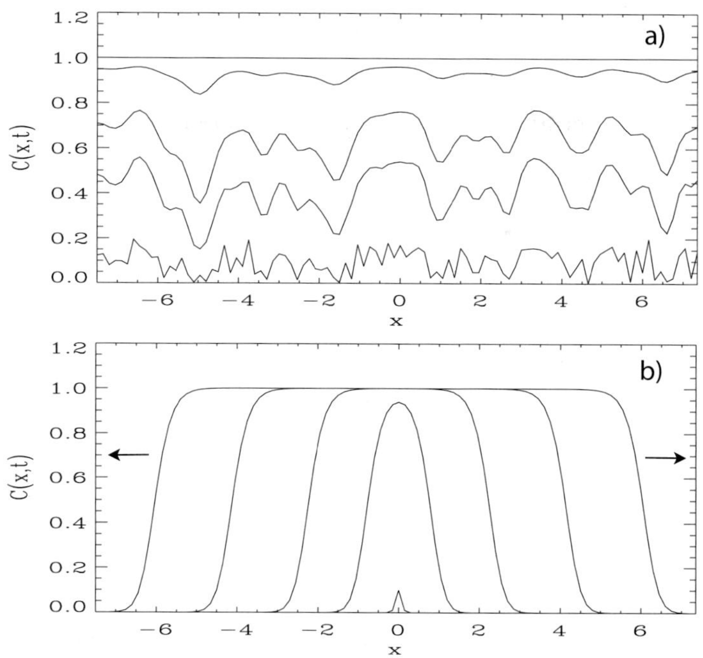

Figure 4.3: Dynamics of the FKKP equation (4.16). $K = 1$, $D = 0.1$, $\mu = 10$. a) The initial condition is distributed over the whole system. Configurations are shown, from lower to highest, at $t = 0$, 0.2, 0.3, 0.5, and 3.5. b) The initial condition is a small perturbation localized in the center of the system. Two clear fronts are formed and advance in opposite directions at constant velocity. From the narrowest to the widest configurations, $t = 0$, 0.7, 1.5, 2.5, and 3.5.

parameter space on which one can expect generically the existence of trajectories representing front solutions. In the present case the intersection turns out to be one-dimensional (if all parameters in (4.19) are positive), so that we expect fronts for a continuous range of values of the parameter v for fixed μ, D and K. Thus the front solution is not unique but there is a one-parameter family of them.

In the present case (4.19), involving a single unknown function g and only first and second order derivatives, these *counting arguments* can be made more explicit and intuitive by writing (4.19) as the mechanical equation of motion of a point particle of mass D at position $g(\xi)$ at time ξ in a potential $U(g)$ and with a friction force proportional to the particle velocity g':

$$D\frac{d^2g}{d\xi^2} = -v\frac{dg}{d\xi} - \frac{dU(g)}{dg} \ . \tag{4.20}$$

The front velocity $v > 0$ plays here the role of a friction coefficient and the potential is

$$U(g) = \mu\frac{g^2}{2} - \frac{\mu}{K}\frac{g^3}{3} \ . \tag{4.21}$$

This potential is plotted in Fig. 4.4. Front solutions are trajectories starting at $g = C_s = K$ with an infinitesimal displacement or initial velocity, and ending at the bottom of the well $g = C_u$ (note that the stability of the states C_s and C_u in this artificial dynamical system evolving in *pseudotime* $\xi = x - vt$ has been inverted with respect to the initial homogeneous states of (4.16), evolving in real time t). It is clear from the mechanical analogy that these trajectories exist for any value of the friction $v > 0$, confirming again the existence of a family of solutions $g_v(\xi)$ parameterized by the value of its speed v. The same conclusion is reached for the more general equation (4.17), with the potential $U(g) = -\int dgF(g)$.

Thus, the question arises: which of these solutions, if any, is the one reached from localized initial conditions in the dynamic process exemplified in Fig. 4.3b. The rigorous result (Kolmogorov et al., 1937; Aronson and Weinberger, 1978) is that, for sufficiently localized initial conditions the travelling front of (4.16) approaches the solution g_v of (4.19) characterized by a speed $v = 2\sqrt{D\mu} \equiv c$. This is exactly the result that is obtained by linearizing (4.16) around the unstable state $C = C_u = 0$, so that it reduces to (4.4) and leads to the asymptotic speed (4.8). The meaning of "sufficiently localized" initial conditions $C(x, t = 0)$ is precisely the same as in the linear case, i.e. decay at $x \to \infty$ faster than $\exp(-a_c x)$, with $a_c = \sqrt{\mu/D}$. Even the approach to the asymptotic velocity is from below and the slow algebraic transient giving the first correction to the long time limit is as in the linear case (4.7), although with a slightly different numerical coefficient in front of the t^{-1} dependence (Ebert and van Saarloos, 2000). $g_v(\xi)$ approaches $C_s = 1$ for $\xi \to -\infty$, $C_u = 0$

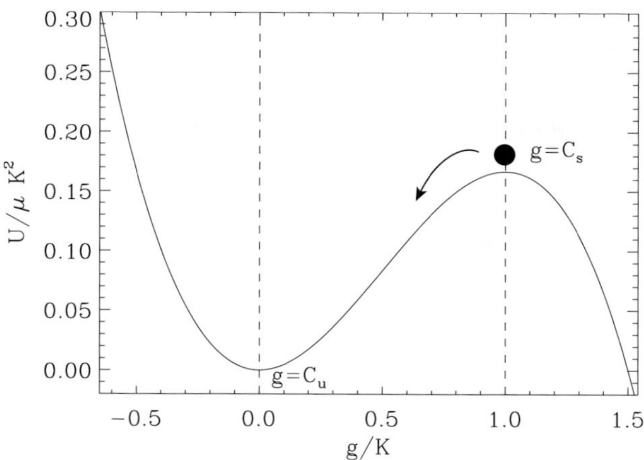

Figure 4.4: The potential (4.21). Fronts g_v correspond to trajectories of a particle which starts at $g = C_s$ with infinitesimal velocity and falls down to $g = C_u$ under friction v. Such solutions exist for any positive value of v.

for $\xi \to \infty$, and the front region containing the transition between the two states has a width $a_c^{-1} = \sqrt{D/\mu}$. The selected speed also coincides with the linear result (4.11) in the case in which the initial steepness is not sufficiently large.

For equations of the type (4.16) or (4.17), the velocity c is also the one below which the point $P_+ = (0,0)$ in the (g, g') phase space ceases to be a stable node and becomes a stable focus. Then, the fronts g_v with $v < c$ present small oscillations around zero at the leading edge, so that they are not valid solutions if $C(x,t)$ represents a positive density, and in this case the selected velocity is the minimum speed corresponding to acceptable solutions. This characterization of c is correct for equations of the form (4.17), but for models consisting of sets of equations, or containing higher order derivatives, or not representing densities, the characterization of c as the one naturally arising from the linear dynamics is the appropriate one.

Thus, it is clear that the front behavior of (4.16) is completely dominated by the linear behavior close to the unstable state C_u. The full solution is only close to C_u in the *leading edge* of the travelling front, so that we can say that the full front is *pulled* by the linear leading edge. This is a behavior found for the advance of a front in a linearly unstable state in a large number of equations (van Saarloos, 2003), but not in all. For example, in the case (4.17) a sufficient condition for the results of the linearization to be valid ($v_f \to c = 2\left(DF'(C_u)\right)^{1/2}$) is that $F(C)/(C - C_u) \leq F'(C_u)$ if $C_u < C < C_s$ (Aronson and Weinberger, 1978). Intuitively this guarantees that the nonlinear terms do not introduce a growth larger than the linear growth. When this condition is violated (as is the case of the function $F(C)$ depicted in Fig. 4.2, right) we can have *pushed* fronts in which the speed is determined by the dynamics occurring in the nonlinear region where $C(x,t)$ is substantially different from C_u (instead of being determined by the leading edge).

In general, solutions of (4.17) of the form $g_v(x - vt)$ with a steepness a will be invaded and replaced by the pulled front originating from the linear perturbations growing at their leading edge if v is smaller than the propagation velocity selected linearly for that steepness (c or $v(a)$). Thus, stable nonlinearly *pushed* fronts can have only a speed $\hat{v}$ larger than the one predicted by the linear approach for that steepness (or be steepest than the linear prediction for its velocity). These arguments can be formalized by studying the linear stability of the front solutions (van Saarloos, 2003; Ebert and van Saarloos, 2000). Determining whether or not a pushed front with these characteristics exists requires global analysis of the full nonlinear problem. Details can be found in van Saarloos (2003). When they exist, they are typically isolated solutions (i.e. not families of solutions) and they are approached exponentially instead of algebraically. In the absence of the required global analysis, a first guess to the problem that turns out to be valid in many situations (van Saarloos, 2003) is to assume that the pushed fronts are absent so that fronts are pulled by the leading edge and the linear analysis gives the correct asymptotic speed. Later analysis or numerical simulation will support or disprove this first guess.

We finally mention that the stability analysis of front solutions mentioned before (van Saarloos, 2003; Ebert and van Saarloos, 2000) reveals that the selected velocity (c or $\hat{v}$ for the pulled or the pushed situation respectively) arising from well localized initial conditions is the *marginally stable* one in the sense that it is the smallest velocity for which the front is stable against invasion by developing localized perturbations.

4.3 Multistability: Fronts advancing on metastable states

Essential to the discussion of the previous Section was the existence of a linearly unstable state. The growth of perturbations on top of it pull the rest of the front structure. We now consider front solutions connecting *linearly stable* states. This happens in situations of multistability, as the ones mentioned in Sect. 3.1.5. If the initial condition is such that initially different parts of the system approach locally one of the possible stable states, competition between them makes the interfaces between the states to move. In this case any front motion should be *pushed* by the nonlinearities since the linear approximation predicts that any perturbation is damped and does not propagate. The situation is illustrated in Fig. 4.5.

We exemplify the discussion with Eq. (4.17) but where now (see Fig. 4.6a) $F(C)$ has three zeros, $C_1 < C_u < C_2$, with $F'(C_1) < 0$, $F'(C_2) < 0$, and $F'(C_u) > 0$, so that C_1 and C_2 are linearly stable and C_u linearly unstable states. As specific examples, we use the dynamics (3.63) or (3.67), written in the form

$$\frac{\partial C}{\partial t} - D\frac{\partial^2 C}{\partial x^2} = F(C) = \mu\,(C - C_1)\,(C_u - C)\,(C - C_2) \ , \qquad (4.22)$$

but most of the discussion does not require to fully specify F. We note that if in the initial condition part of the system is close to the C_u unstable state and part close to only one of the two stable states, a front in which the stable state advances on the unstable one will be formed, multistability will be irrelevant, and the situation will be similar to the one discussed in Sect. 4.2. The new situation appears when the initial condition brings the system locally close to both stable states, as in Fig. 4.5, so that competition between the stable states becomes relevant.

We can search for solutions of the form $C(x,t) = g(x - vt) \equiv g(\xi)$, so that the equation for $g(\xi)$ is

$$Dg''(\xi) + vg'(\xi) + F(g(\xi)) = 0 \ . \qquad (4.23)$$

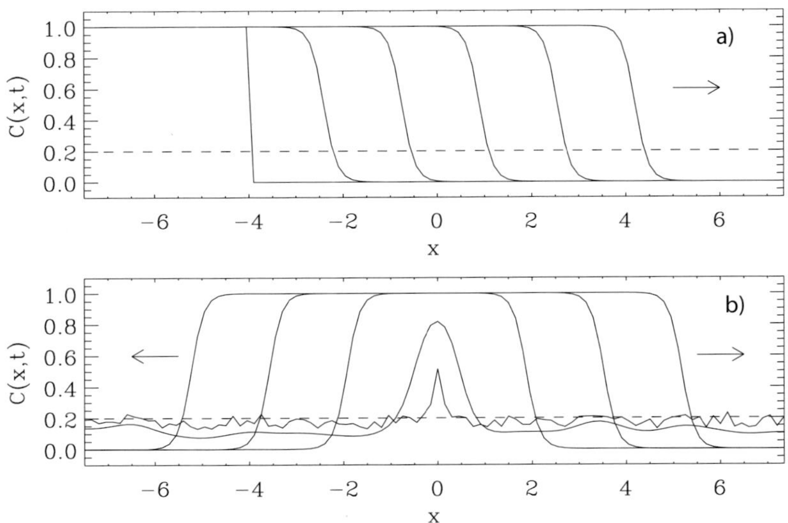

Figure 4.5: Solutions of Eq. (4.22) with $\mu = 10$, $D = 0.1$, $C_1 = 0$, $C_u = 0.2$, and $C_2 = 1$. Horizontal dashed lines are at $C = C_u$. a) The initial condition is a sharp jump between the two stable states, $C = 0$ and $C = 1$. A front is formed and moves towards the right at constant speed, so that state $C = 1$ invades and replaces the $C = 0$ one. The configurations shown correspond, from left to right, to $t = 0, 4, 8, 12, 16$, and 20. b) The initial condition is an irregular function with its central part higher than the lateral ones. Times shown are, from the interior to the exterior configurations, $t = 0, 1, 4, 8$, and 12. Regions of $C = 1$ and $C = 0$ are formed soon, after which the $C = 1$ invades the other one with the propagation of two fronts similar to the one in a).

Counting arguments dealing with the manifolds of the fixed points of (4.23) now suggest that solutions of the nonlinear eigenvalue problem (4.23) are isolated. This is confirmed, as in the case discussed in the previous Section, by realizing that (4.23) can be written as (4.20), which represents the motion of a point particle of mass D at position $g(\xi)$ at time ξ in a potential $U(g) = -\int dg F(g)$ under a damping v. $U(g)$ in this case is plotted in Fig. 4.6 (compare with Fig. 4.4). Now the front solution is a heteroclinic trajectory between the two maxima of U. Only fine tuning of the damping v to a particular value allows such trajectory. If $v > 0$, this is possible only if the state at

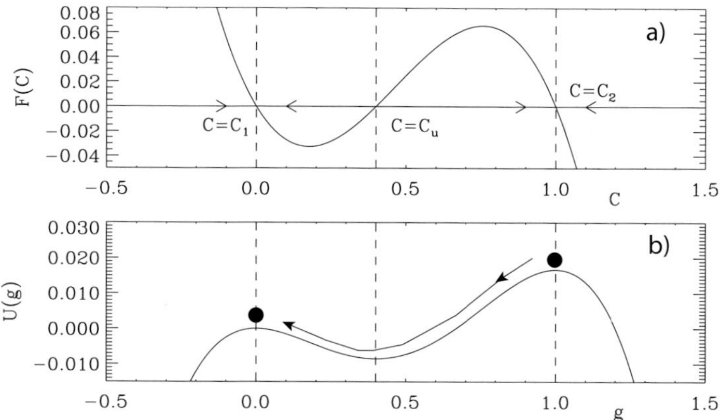

Figure 4.6: a) The function $F(C)$ in (4.22). The three zeros are at $C_1 = 0$, $C_u = 0.4$, and $C_2 = 1$. The arrows indicate the direction of motion of the dynamical system $\dot{C} = F(C)$ in each C-interval. C_u is a unstable fixed point and C_1 and C_2 are linearly stable ones. b) The potential $U(g) = -\int dg F(g)$. Front solutions of (4.20) joining the two stable states correspond to trajectories of a particle which starts at $g = C_2$ with infinitesimal velocity, falls down under friction v and stops precisely at C_1. Such solutions exist just for a particular value of v.

the initial time, say $g(\xi = -\infty)$ is the one with the higher value of U (C_2 in Fig. 4.6). Another way to establish the sign of v, and thus the direction of the front motion is to multiply Eq. (4.23) by $g'(\xi)$, and integrate over ξ on the full real line. Since $2g''g' = \left((g')^2\right)'$, $g'(-\infty) = g'(\infty) = 0$, and assuming as in Fig. 4.5 that $g(-\infty) = C_2$ and $g(\infty) = C_1$, we have

$$\int_{-\infty}^{\infty} d\xi g'(\xi) F(g(\xi)) = \int_{C_2}^{C_1} dg F(g) = U(C_2) - U(C_1) , \qquad (4.24)$$

and

$$v = \frac{U(C_2) - U(C_1)}{\sigma} \qquad (4.25)$$

with $\sigma = \int_{-\infty}^{\infty} (g')^2 d\xi$. This last formula is not a closed expression for v, since to calculate σ one has still to solve the nonlinear eigenvalue problem for g, (4.23), but since $\sigma > 0$ it confirms that the sign of the front speed v will be such that the state with higher value of U invades the one with lower value. This is equivalent to state that the direction of motion is determined by the sign of the area under the curve $F(g)$ (see Eq. (4.24)). We can say that the state which is invaded is only *metastable*, whereas the one winning finally the competition is *globally stable*. The front becomes steady ($v = 0$) when $U(C_1) = U(C_2)$ or, equivalently, when the lobes in Fig. 4.6a enclosed between the horizontal axis and the upper and lower parts of $F(C)$ are equal. This is analogous to the Maxwell construction in thermodynamics to establish the condition for phase equilibrium, with U playing the role of the negative of a free energy.

These ideas can be made more concrete for the simple case (4.22), for which explicit exact solutions are available. For instance, one can check by substitution that an exact solution of (4.23) joining $g(\xi = -\infty) = C_2$ with $g(\xi = \infty) = C_1$ is:

$$C(x,t) = g(\xi) = \frac{C_2 + C_1 e^{a(C_2-C_1)(\xi-\xi_0)}}{1 + e^{a(C_2-C_1)(\xi-\xi_0)}} . \qquad (4.26)$$

ξ_0 fixes the arbitrary initial location of the front, the front width is a^{-1}, with $a = \sqrt{\mu/(2D)}$, $\xi = x - \hat{v}t$, and

$$\hat{v} = \sqrt{2\mu D} \left(\frac{C_1 + C_2}{2} - C_u \right) . \qquad (4.27)$$

For the parameters in Fig. 4.5, $C_u < (C_1+C_2)/2$, so that the globally stable C_2 invades the metastable C_1, in agreement with Eq. (4.25) and Fig. 4.6, where we see that $U(C_1) < U(C_2)$. Function (4.26) reproduces very well the shape of the front in Fig. 4.5a and of each of the two front regions in Fig. 4.5b at late times.

The simple structure of (4.22) allows additional analytic solutions (Sanati and Saxena, 1998) which may be representative of other bistable systems. For example, the potential analogy in Fig. 4.6 shows that, if dissipation is absent ($v = 0$, which means that we are looking for a steady solution) there is generically a trajectory which starts at the metastable C_1, rolls down passing C_u in the direction of C_2 but, before arriving, returns back and stops again precisely at C_1. Such solution, homoclinic to C_1, has the shape of a pulse as shown in Fig. 4.7. Its analytic expression is (Sanati and Saxena, 1998):

$$C_p(x) = C_1 + C_+ - \frac{C_+ - C_-}{1 - \frac{C_-}{C_+} \tanh^2\left(\frac{x - x_0}{w_p}\right)} \tag{4.28}$$

with

$$C_\pm = \frac{2}{3}(C_2 + C_u - 2C_1)$$

$$\pm \left(\frac{4}{9}(C_2 + C_u - 2C_1)^2 - 2(C_2 - C_1)(C_u - C_1)\right)^{\frac{1}{2}} . \tag{4.29}$$

$C_1 + C_+$ is the maximum pulse concentration, x_0 is the arbitrary center location, and the width is

$$w_p = 2\sqrt{\frac{D}{\mu(C_u - C_1)(C_2 - C_1)}} . \tag{4.30}$$

This is a steady solution, but it is dynamically unstable: any small perturbation makes the system to leave it. Thus it does not play a role in dynamics such as the ones shown in Fig. 4.5, but it will allow understanding of some results in Chapter 7.

We stress that, although the particular expressions (4.26)-(4.27) or (4.28)-(4.30) are only valid for the particular cubic nonlinearity in (4.22), the existence and general shape of these solutions remains valid for any function $F(C)$ having three zeros as in Fig. 4.6. We should also say however that the particle-in-a-potential analogy used above to establish these generic results is only possible for dynamics involving a single species in one-dimensional space under diffusive dispersion, or situations reducible to that. It is beyond our goal to describe fully other bistable situations which may involve for example multiple species or nonlinear diffusion (see for example Murray (1993)). For the case of two spatial dimensions and a single species we mention that one possible situation is that of a flat front separating two regions containing each of the two stable states. In this case the previous results will apply since there is only one relevant direction which plays the role of x in the above analysis, the one perpendicular to the front. The speed of front propagation v will be

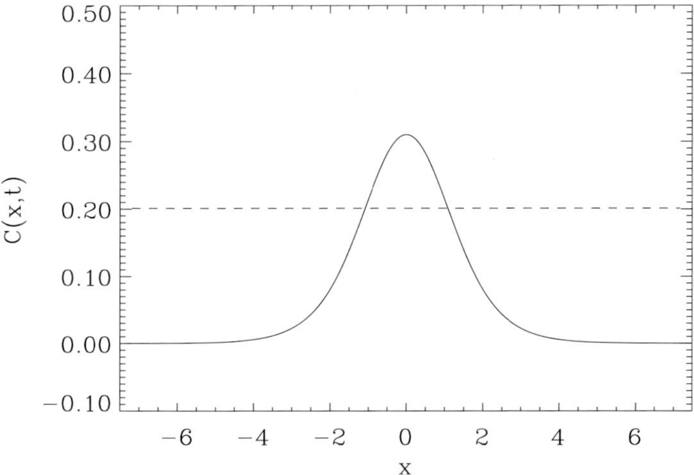

Figure 4.7: The unstable steady pulse solution Eqs. (4.28)-(4.30), centered at $x_0 = 0$, for the same parameter values as Fig. 4.5. Dashed horizontal line is at $C = C_u$.

the one given by the one-dimensional analysis, say v_1. If the front becomes curved, the curvature κ modifies this unique speed of propagation normal to the front in a way that, in the limit of thin fronts, becomes (Allen and Cahn, 1979; Bray, 1994)

$$v = v_1 + D\kappa \ . \tag{4.31}$$

The sign in the curvature term in (4.31) is such that the state in the convex part of the front pushes into the concave one. Thus, depending on the geometry, this curvature correction can help or go against the term v_1 determined by the relative stability of the two competing states. In any case the κ term tends to flatten localized curvature perturbations in moving fronts, so that asymptotically the flat front and thus the one-dimensional front speed is the relevant one. The methodology to extend Eq. (4.31) to multispecies situations is available (see e.g. Gomila et al. (2007)).

4.4   Excitable waves

As mentioned in Chapter 3, the type of excitable behavior discussed
there may be considered as arising from a quasi-bistable dynamics in
which one of the involved states, the excited one, is not really stable
but lasts only for a finite time. Thus excitable diffusive systems have
some similarities with bistable ones, but present an additional level
of complexity.

We consider in this section, as an example, the $P\text{-}Z$ dynamics of
Eqs. (3.77)-(3.78), but with an additional diffusion terms:

$$\frac{dP}{dt} = r\left(1 - \frac{P}{K}\right)P - a\frac{P^2}{P_0^2 + P^2}Z + D\nabla^2 P \qquad (4.32)$$

$$\frac{dZ}{dt} = b\frac{P^2}{P_0^2 + P^2}Z - cZ + D\nabla^2 Z . \qquad (4.33)$$

$P(x,t)$ is the phytoplankton concentration and $Z(x,t)$ the zooplankton one. To avoid unnecessary complications we assume that the two
species disperse with the same diffusion coefficient D.

We show in Fig. 4.8 simulations of the model in the one-dimensional
situation in the excitable regime. Other excitable systems behave
qualitatively in the same way. The initial condition is a phytoplankton pulse on top of the $P\text{-}Z$ steady steady state. When the perturbation is larger than the excitation threshold, the excited state
develops and their flanks begin to expand into the unexcited state as
the fronts in the bistable case of Fig. 4.5b. But the slow dynamics
of the Z variable finally manifests itself by recalling that the excited
state is not a stable one: The middle of the excited patch returns
to the unexcited stable state, so that the pair of counterpropagating fronts becomes a pair of counterpropagating pulses. Each pulse
can be considered to be a bound state of two fronts, the leading
one from the stable to the excited state, and the trailing one, from
the excited back to the stable. In biological terms the phenomenon
is clear: when positively perturbed, the fast reproduction ability of
phytoplankton allows it to escape the grazing control from zooplankton so that it blooms and begins to invade neighboring areas. Some
time later zooplankton has grown where there is more phytoplankton and consumes it, but always lagging behind, so that the pulse

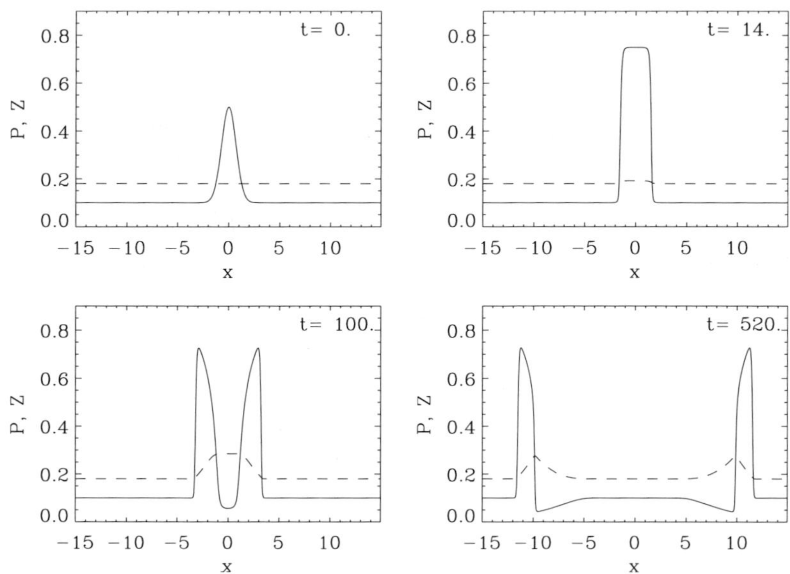

Figure 4.8: Plankton excitable waves from system (4.32)-(4.33) for $r = K = a = 1$, $P_0 = 0.1$, $b = 10^{-2}$, $c = 0.5b$, and $D = 10^{-3}$, in the one-dimensional case. Solid line is $P(x,t)$, dashed $Z(x,t)$. From a localized perturbation to the equilibrium state, the excited state develops and splits in two pulses propagating in opposite directions.

structure is a phytoplankton patch whose rear part is being grazed by zooplankton. In this context the *activator* and *inhibitor* variables of Chapter 3 are conveniently called *propagator* and *controller*, respectively (Fife, 1984). The pulse speed is controlled by the leading edge and details of the shape of the front may be found by matching the speed of the trailing front to the leading one (Meron, 1992; Murray, 1993). Note that the trailing front has two parts, the fast one in which P drops suddenly to values on the nullcline ($F(P, Z) = 0$ in Eq. (3.77)), and a smoother one in which both P and Z slowly recover along the nullcline and approach the fixed point. This last part is called *refractory* and has the interesting property of being relatively insensitive to perturbations, since the P-Z concentrations are still not close enough to the excitable values. Sufficiently far from the pulse the values of P-Z are finally at the excitable fixed point,

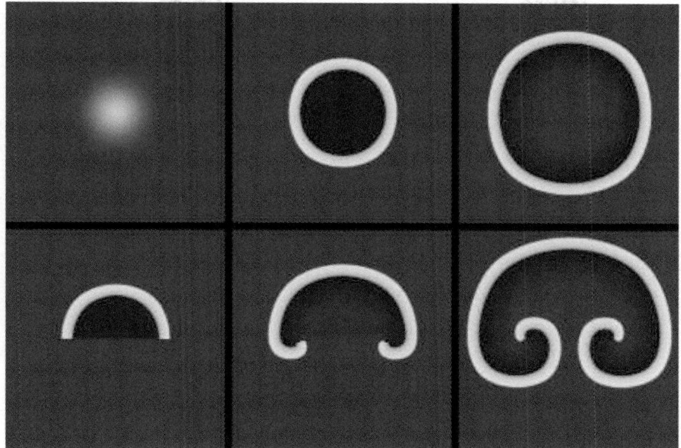

Figure 4.9: Phytoplankton $(P(x,t))$ concentrations in excitable waves from system (4.32)-(4.33) for $r = K = a = 1$, $P_0 = 0.1$, $b = 10^{-2}$, $c = 0.5b$, and $D = 5 \times 10^{-3}$, in two-dimensions. White is $P = 0.8$ and black is $P = 0$. Upper row: From a localized perturbation to the equilibrium state, the excited state develops and gives rise to a target wave expanding outwards. Times shown are, from left to right, $t = 0$, 150, and 350. Lower row: The initial condition is the middle state in the upper row, but with half of the system reset to the fixed point ($P = 0.1$, $Z = 0.18$). Spirals develop from the excited pulse endpoints. The snapshots are taken at the same times from initial condition as above.

where they can be readily excited by new small (but superthreshold) perturbations.

When the same dynamics occurs in an isotropic two-dimensional medium, the pair of counterpropagating pulses becomes an outwards propagating ring, or *target wave*, formed by circular leading and trailing fronts (see Fig. 4.9). The speed of the fronts depends on the curvature, as it was the case in Eq. (4.31). An important consequence of this dependence is that, if the pulse becomes broken (e.g. by a perturbation) so that the ends of the front have a large curvature, strong deformation occurs. The lower row of Fig. 4.9 shows that the pulse typically winds around itself forming spiral patterns. When target waves travel in inhomogeneous or in stirred media they are likely to break and then spirals are observed.

Figure 4.10: Target and spiral waves in the ferroin-catalized BZ reaction. Right photograph courtesy of V. Pérez-Muñuzuri.

Excitable behavior in general and target and spiral patterns in particular have been observed in a huge variety of chemical and biological systems, including electric potential in hearth (Davidenko et al., 1992) or brain (Shibata and Bureš, 1974) tissues, calcium signals in eggs (Camacho and Lechleiter, 1993), aggregation of slime mold cells (Newell, 1983), and of course in excitable chemical reactions and related systems (Epstein and Pojman, 1998; Meron, 1992; Murray, 1993; Cross and Hohenberg, 1993). Figure 4.10 shows target and spiral patterns in the BZ reaction. Note that periodic trains of targets are not expected in an excitable system. They are thought to originate from dust particles which locally bring the medium into the oscillating regime.

There are also observations of plankton waves in the ocean (Srokosz et al., 2004; Martin, 2003), although their behavior is expected to be strongly influenced by the ambient flow (Abraham et al., 2000; Neufeld et al., 2002a) as discussed later in Chapter 7. It is however difficult to assess if the mechanism behind these waves is really excitability. In fact a Fisher-like wave of phytoplankton expanding into uncolonized waters, followed by a zooplankton predation wave (Murray, 1993; Dubois, 1975a,b) will have a similar aspect. Also, oscillatory predator-prey dynamics such as in (4.32)-(4.33) for other parameter values or when replacing the functional response by a Holling-type-II displays waves for some initial conditions, among many other possible dynamical structures. But they are generically wave trains, not isolated pulses.

We finally mention that wave propagation arises from diffusion, but that the excitable behavior can give rise to interesting spatiotemporal phenomena even without diffusion. For example, Matthews and Brindley (1997) show for model (4.32)-(4.33) with $D = 0$ that if there is a spatially localized forcing (a point source of nutrients, a local change in temperature, etc.), a localized excitation and recovery plankton pulse will occur, which may be a mechanism for localized (i.e. non-propagating) plankton blooms.

4.5 Turing diffusive instabilities

Diffusion is expected to homogenize concentrations and thus it is usually thought as a stabilizing effect. Turing (1952) identified a first example showing that this is not always the case, and proposed it as a possible mechanism for morphogenesis. The ingredients are an activator-inhibitor dynamics and diffusion. But the important assumption is that the inhibitor diffuses faster than the activator. Essentially one has "local activation and lateral inhibition": the activator species (e.g. autocatalytic) tends to grow at a rate such that it can not be controlled by the inhibitor which consumes it. But because of its relatively small diffusivity, the growth remains in place. The inhibitor can diffuse further, and control the growth where the inhibitor is just arriving. Since there can be no inhibitor too far from the activator, at longer distances the activator is again free to grow and the elementary pattern of activator surrounded by inhibitor repeats periodically. Thus, periodic arrangements of activator and inhibitor patches are expected to arise as a consequence of this dynamics. In two dimensions this leads typically to hexagonal patterns, but stripes can also appear (see Fig. 4.11). Additional details of this mechanism which leads to the instability of homogeneous concentrations and to the appearance of steady periodic patterns can be found in Walgraef (1996), Murray (1993), or Epstein and Pojman (1998).

The Turing instability was proposed to describe qualitatively natural phenomena such as animal skin patterns (Murray, 1993), but it took close to 40 years to be reproduced experimentally in a well con-

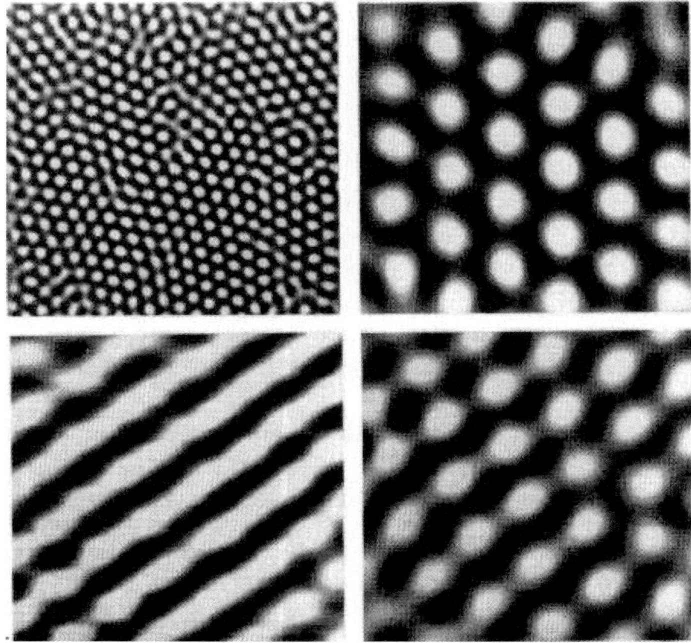

Figure 4.11: Turing patterns observed in a CIMA reaction. Light and dark colors are the two states of an indicator for the I_3^- concentration. From Ouyang and Swinney (1991). Reprinted by permission from Macmillan Publishers Ltd: Nature 352, 610-612, copyright 1991.

trolled chemical system, the CIMA reaction (Castets et al., 1990). The difficulty was to obtain the needed large difference between diffusion coefficients, which for usual molecules in aqueous solution are of the same order. The problem was solved by running the reaction in a gel and binding some of the reactants to long starch molecules, which are effectively immobile in the gel medium. Figure 4.11 shows different steady patterns obtained for the CIMA reaction in a gel medium.

Now a number of chemical and biological systems (Epstein and Pojman, 1998; Murray, 1993) are known to display Turing patterns. The mechanism has also been invoked as possible source of plankton patchiness (Levin and Segel, 1976). Although diffusion coefficients of planktonic organisms have roughly the same value for both predator (zooplankton) and prey (phytoplankton), the former usually has a larger vertical mobility. If there is some vertical shear in the water column, the organisms with larger vertical mobility will experience a more variable velocity field, and thus they will be subjected to faster dispersion by the mechanism of shear dispersion (Sect. 2.2.2). If one models such enhanced dispersion by an effective diffusion coefficient, it would be larger for zooplankton than for phytoplankton, and then the Turing mechanism may be at work. The analysis in Matthews and Brindley (1997), however, suggests that for generic $P - Z$ models the Turing mechanism is not likely to occur in the relevant range of parameters. But this may not be the case for related but modified models (Bartomeus et al., 2001; Alonso et al., 2002). In addition, related mechanisms, that would lead also to spatially periodic plankton distributions, may appear in the presence of differential motion between predator and prey (Rovinsky et al., 1997). It is difficult to think, however, that the regular structures characteristic of Turing patterns will resist the motions of the turbulent ocean: they will be greatly influenced by flow. It should be remembered that Turing patterns were only clearly observed in chemical systems when aqueous solutions were replaced by immobile gels.

4.6   Oscillatory media and beyond

All the complex behavior described so far in this Chapter arises from the diffusive coupling of the local dynamics which in the homogeneous case have simple fixed points as asymptotic states. If the local dynamics becomes more complex, the range of possible dynamic behavior in the presence of diffusion becomes practically unlimited. It is clear that coupling chaotic subsystems could produce an extremely rich dynamics. But even the case of periodic local dynamics does so. Diffusively coupled chemical or biological oscillators may become synchronized (Pikovsky et al., 2003), or rather additional instabilities may arise from the spatial coupling. This may produce target waves, spiral patterns, front instabilities and several different types of spatiotemporal chaos or phase turbulence (Kuramoto, 1984).

Whereas it is impossible to properly summarize all these complex dynamic behavior, there is a situation in which it becomes rather universal and can be described in a unified way: when all the local oscillations are close to a Hopf bifurcation. In this case it is possible to derive a general equation for the complex envelope $\psi(\mathbf{x}, t)$ of the weakly nonlinear oscillations, known as the Complex Ginzburg-Landau equation (Kuramoto, 1984; Cross and Hohenberg, 1993; Aranson and Kramer, 2002), which in proper dimensionless units reads:

$$\frac{\partial \psi}{\partial t} = \psi + (1 + i\alpha)\nabla^2\psi - (1 + i\beta)|\psi|^2\psi \qquad (4.34)$$

α measures the changes in local oscillation frequency by spatial inhomogeneities, and β its dependence on the oscillation amplitude. Regions in the parameter space (α, β) displaying qualitatively different behavior (stable phase waves, phase turbulence, defect turbulence leading to spiral chaos, intermittency, etc.) have been mapped both in one (Shraiman et al., 1992; Chaté, 1994, 1995) and in two (Chaté and Manneville, 1996) dimensions, which should serve as a guide for the behavior of diffusive oscillatory media close to the onset of oscillation.

Chapter 5

Fast Binary Reactions and the Lamellar Approach

As a first case study to consider the impact of advection processes on reaction-diffusion dynamics (thus leading to the full reaction-diffusion-advection problem) we address here the case of a binary reaction. The limit of a fast reaction (as compared with diffusion) becomes simple to analyze, and the "filament" or "lamellar" approach developed in Sect. 2.7.1 is a very appropriate tool to understand the dynamics in simple geometries and to interpret observations in more complex ones.

5.1   Lamellar reacting models

The idea of focussing on the transverse profile of filaments (see Sect. 2.7.1) has a long history, specially in the chemical engineering and combustion contexts (Carrier et al., 1975; Ranz, 1979; Ottino, 1994; Bish and Dahm, 1995), where it appears under the names of *lamellar models* or *stretch models*. More recently it has been explicitly used by Neufeld (2001), Neufeld et al. (2002b), McLeod et al. (2002), Neufeld et al. (2002c), Szalai et al. (2003), Hernández-García et al. (2003), or Cox (2004) and is one of the ideas behind the approaches used in papers such as Jimenez and Martel (1991), Wonhas and Vassilicos (2002), or the ones reviewed in Tél et al. (2005).

In a reaction-diffusion-advection formulation, the model considers a one-dimensional slice transverse to concentration filaments, and models its evolution by an equation of the type

$$\frac{\partial C_i}{\partial t} - \lambda(t)x\frac{\partial C_i}{\partial x} = D\frac{\partial^2 C_i}{\partial x^2} + R_i\left(C_1,\ldots,C_N\right). \qquad (5.1)$$

Equation (4.13) is a particular case. $\{C_i\}$ is a set of concentrations interacting through the reaction terms R_i and $v_x = \lambda(t)x$ is the, possibly time-dependent, transverse velocity field pointing towards the center of the filament located at $x = 0$. As in Sect. 2.7.1 this is to be understood as a local Lagrangian description.

The change of variables (2.92) makes here also the advection term in (5.1) to disappear (we stress that equality of the diffusion coefficients of all species is needed; see Kiss et al. (2003b) for an example in which this condition is not fulfilled). In the new coordinates the equations for the concentrations $\tilde{C}_i(\xi,\tau)$ defined by $\tilde{C}_i(\xi(x,t),\tau(t)) = C_i(x,t)$ read

$$\frac{\partial \tilde{C}_i}{\partial \tau} = \frac{\partial^2 \tilde{C}_i}{\partial \xi^2} + \tilde{R}_i\,. \qquad (5.2)$$

The advection terms have disappeared, at the expense of introducing time-dependent reaction rates: $\tilde{R}_i \equiv s(t)^2 R_i/D$. Thus, usually this approach can be used to obtain analytic solutions only when the reaction term can be neglected or eliminated by additional changes of variables. This is the case for example with Eq. (4.13), since the reaction term $\mu(t)P$ can be eliminated by considering the equation for $C(x,t)$ defined by $P(x,t) = C(x,t)\exp\left(\int_0^t \mu(u)du\right)$. All the solutions discussed in Sect. 2.7.1 can be applied to that case. In the following we describe another situation in which one can deal with the reaction term.

5.2   Fast binary reactions in simple flows

Consider the binary chemical reaction $A + B \to C$. The reaction-diffusion-advection equations read, in the case of equal diffusion coefficients ($A = A(\mathbf{x}, t)$, $B = B(\mathbf{x}, t)$, and $C = C(\mathbf{x}, t)$ are the concentrations of the respective chemicals):

$$\frac{\partial A}{\partial t} + \mathbf{v} \cdot \nabla A = D\nabla^2 A - kAB \tag{5.3}$$

$$\frac{\partial B}{\partial t} + \mathbf{v} \cdot \nabla B = D\nabla^2 B - kAB \tag{5.4}$$

$$\frac{\partial C}{\partial t} + \mathbf{v} \cdot \nabla C = D\nabla^2 C + kAB \ . \tag{5.5}$$

The symmetry of this system allows to find conservation laws analogous to the spatially independent case (3.8). For example, the equation for the quantity $\phi(\mathbf{x}, t) \equiv A(\mathbf{x}, t) - B(\mathbf{x}, t)$, analogous to Q_3 in (3.8) is:

$$\frac{\partial \phi}{\partial t} + \mathbf{v} \cdot \nabla \phi = D\nabla^2 \phi \tag{5.6}$$

giving a standard diffusion-advection problem. After solving Eq. (5.6), one of the concentrations, say A, can be recovered from (5.3) by using $B = A - \phi$:

$$\frac{\partial A}{\partial t} + \mathbf{v} \cdot \nabla A - D\nabla^2 A = -kA\left(A - \phi\right) \tag{5.7}$$

which is analogous to (3.9).

At this point, the problem can be simplified to a non-reacting one in the limit of *fast reaction*: When $k \to \infty$ (or Da $\to \infty$), the right hand side of (5.7) is very large, leading to a very fast change everywhere except at the tiny regions where A and B are both small. Thus we can assume the last term in (5.7) to be zero after a short transient everywhere except at the interfaces between A and B. At any point of space, either $A = 0$ (and then $B = -\phi$, which can only happen where $\phi < 0$), or $A = \phi$ (which can only happen where $\phi > 0$, and $B = 0$). The reactant in minority at each place is immediately consumed and A and B can not coexist. Since k is infinity, however, the reaction can still occur at the interface between A and B. After entering this regime, the reaction is controlled just by diffusion towards the reactive interfaces, and affected by their advective motion. The quantity $T(\mathbf{x}, t) \equiv A(\mathbf{x}, t) + B(\mathbf{x}, t) + 2C(\mathbf{x}, t)$ also satisfies a simple advection-diffusion equation without reaction term, such as

(5.6). If the initial condition is such that $T(\mathbf{x}, t = 0)$ is homogeneous in space (typical conditions in which initially the reactants are well separated and with the same concentration each, and the product is absent, satisfy this), then it will remain homogeneous and constant under evolution: $T(\mathbf{x}, t) = T \; \forall \, \mathbf{x}, t$. Under these assumptions $\phi(\mathbf{x}, t)$, the solution of (5.6), completely determines all the concentrations at each time and spatial location:

$$\text{where} \quad \phi > 0 \; : \quad A = \phi \, , \, B = 0 \, , \qquad C = \frac{1}{2}(T - \phi) \qquad (5.8)$$

$$\text{where} \quad \phi < 0 \; : \quad A = 0 \, , \, B = -\phi \, , \qquad C = \frac{1}{2}(T + \phi) \, . \qquad (5.9)$$

In the context of combustion in non-premixed flames, this is called the Burke-Schumann solution. If the initial condition is such that $T(\mathbf{x}, t = 0)$ is not homogeneous in space, solving its advection-diffusion equation to obtain $T(\mathbf{x}, t)$ is needed in addition to solving the one for $\phi(\mathbf{x}, t)$ before using (5.8)-(5.9) to obtain $C(\mathbf{x}, t)$.

We now solve the above problem in a pure strain flow $\mathbf{v}(\mathbf{x}) = (-\lambda x, \lambda y)$ in which initially $A = 1$ in $x > 0$ and 0 elsewhere, and $B = 1 - A$, with these concentrations maintained at all time at $\pm\infty$. The initial configuration is shown in Fig. 5.1 (left). This would correspond to a fast binary reaction occurring at the meeting of two counterpropagating streams of reactants, such as an oxidizer and the fuel in combustion, or an acid and a base in a neutralization reaction. Assuming rapid homogenization along the diverging direction of the flow, y, so that $\phi(x, y, t) \approx \phi(x, t)$, Eq. (5.6) for the dependence of ϕ along the converging direction x becomes:

$$\frac{\partial \phi}{\partial t} - \lambda \frac{\partial \phi}{\partial x} = D \frac{\partial^2 \phi}{\partial x^2} \qquad (5.10)$$

with $\phi(x, t = 0) = \text{sgn}(x)$, $\phi(x \pm \infty, t) = \pm 1$. This problem can be solved by using the change of coordinates (2.97) which converts the equation into the diffusion equation $\partial_\tau \tilde{\phi} = \partial_\xi^2 \tilde{\phi}$. The solution with the corresponding boundary conditions is

$$\tilde{\phi}(\xi, \tau) = \text{erf}\left(\frac{\xi}{2\tau^{1/2}}\right), \qquad (5.11)$$

from which the exact time-dependent solution can be obtained by using again (2.97). At long times $\xi/(2\tau^{1/2}) \to (\lambda x^2/2D)^{1/2}$ and a steady solution is reached in the original x, t variables. From this and relations (2.97) and (5.8)-(5.9) the steady concentration profiles are

$$A(x) = 0 \ , \ B(x) = -\mathrm{erf}\left(x\sqrt{\frac{\lambda}{2D}} \right) ,$$

$$C(x) = \frac{1}{2}\left(1 + \mathrm{erf}\left(x\sqrt{\frac{\lambda}{2D}} \right) \right) , \ \text{if } x < 0 \qquad (5.12)$$

and

$$A(x) = \mathrm{erf}\left(x\sqrt{\frac{\lambda}{2D}} \right) , \ B(x) = 0 ,$$

$$C(x) = \frac{1}{2}\left(1 - \mathrm{erf}\left(x\sqrt{\frac{\lambda}{2D}} \right) \right) , \ \text{if } x > 0 \ . \qquad (5.13)$$

These solutions are plotted in Fig. 5.1. The reaction rate k does not enter the above expressions in the $k \to \infty$ limit considered here. We see that a sharp interface between A and B, located at $x = 0$, remains at all times. Also the product C remains concentrated around this interface, forming, in the complete two-dimensional system, a strip of width $\sqrt{2D/\lambda} \sim \mathrm{Pe}^{-1/2}$. The amount of product on a portion of this strip of length L along the transverse direction y is given by $M \equiv \int_L dy \int_{-\infty}^{\infty} dx C(x)dx = LC_T$, where the amount of product per unit of transverse interface is

$$C_T \equiv \int_{-\infty}^{\infty} C(x)dx = \sqrt{\frac{2D}{\pi\lambda}}. \qquad (5.14)$$

We can also calculate the total rate of product creation per unit of time in the system:

$$\mathcal{P} \equiv \int dxdy \ kAB. \qquad (5.15)$$

It would be delicate to calculate this directly from its definition, because of the presence of the infinite reaction rate k. But one can

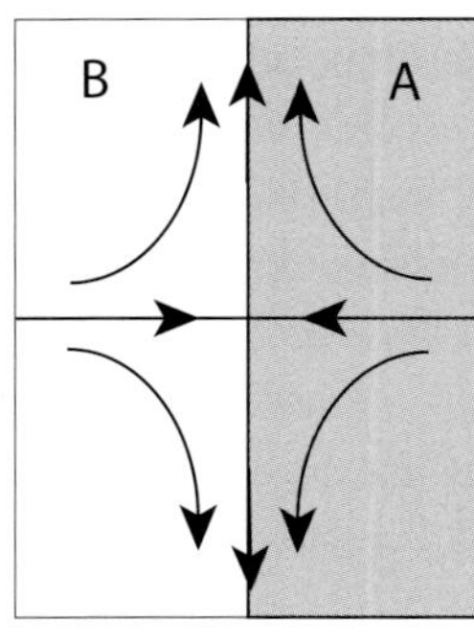

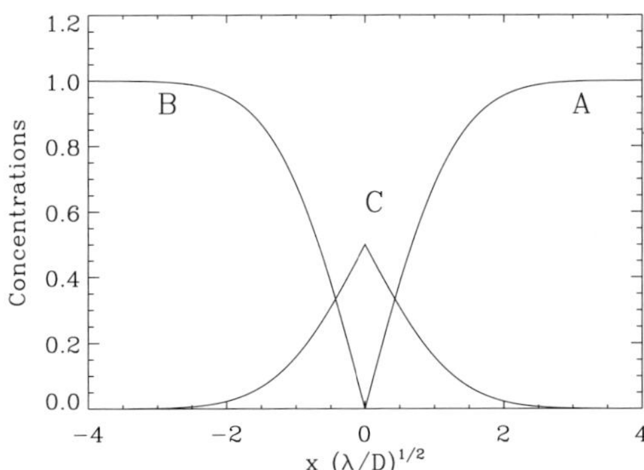

Figure 5.1: The binary $A + B \rightarrow C$ reaction in the fast reaction limit under a straining flow. Left, the initial reactant configuration, with flow lines indicated. Right, the steady state concentrations from Eqs. (5.12)-(5.13)

consider instead the equation (5.5) in the form (5.1) for the product generation in this flow:

$$\frac{\partial C}{\partial t} = \lambda x \frac{\partial C}{\partial x} + D \frac{\partial^2 C}{\partial x^2} + kAB = -\lambda C + \frac{\partial}{\partial x}\left(\lambda x C + D \frac{\partial C}{\partial x}\right) + kAB, \tag{5.16}$$

which can be integrated to obtain the rate of product generation per unit interface length, or *production*:

$$p \equiv \frac{\mathcal{P}}{L} = \dot{C}_T + \lambda C_T \ . \tag{5.17}$$

We note that this last equation is completely general for the production of a chemical which remains in a strip aligned with the y direction. In our case $\dot{C}_T = 0$, since the profile is stationary, and using (5.14):

$$p = \lambda C_T = \sqrt{2\lambda D/\pi} \ . \tag{5.18}$$

In this fast reaction $k \rightarrow \infty$ limit, the reaction is limited by diffusion so that it proceeds faster at larger diffusion coefficients. But the strain λ also contributes to the increase of the production. Note that, despite this increase of production rate with λ, the amount of product per unit of length diminishes (Eq. (5.14)).

The same approach can be used to find the interface structure for different kinds of time-dependent strain. For example, both the shear or the vortex flows described in Sect. 2.7.1 lead to interfaces described by (5.11) with $\xi/(2\tau^{1/2}) \to x/\sqrt{4Dt/3}$ at long times. In these cases the interface region, aligned with the shear flow, widens diffusively in time. We have $C_T \sim \sqrt{Dt}$ and the production per unit length $p = \dot{C}_T(t) + \lambda(t)C_T(t) \sim (D/t)^{1/2}$ decreases in time. At any fixed time both the interface width and the production behave as $\mathrm{Pe}^{-1/2}$, as in the pure strain case.

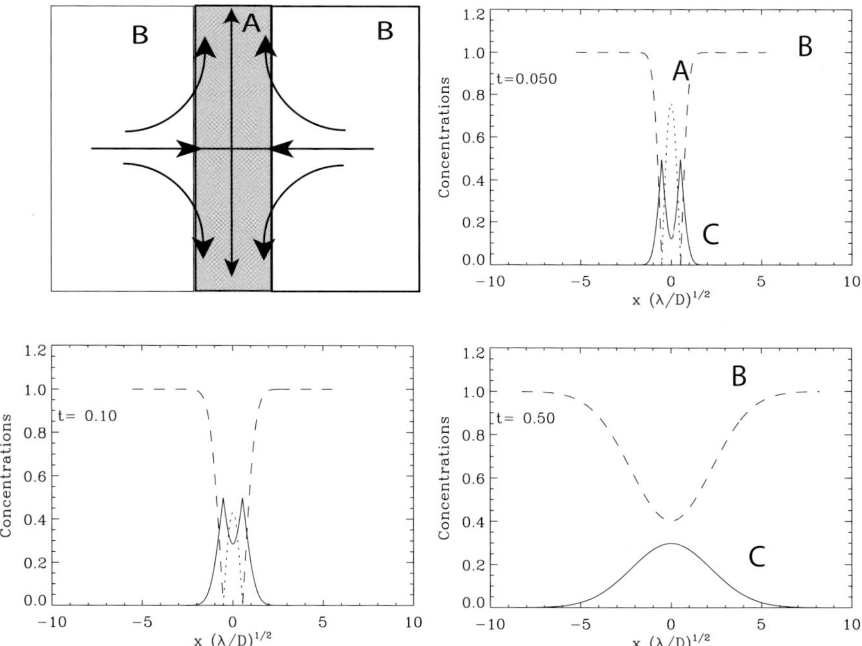

Figure 5.2: Same as Fig. 5.1 but for the initial configuration displayed in the upper-left panel: There is a central A region surrounded by the B reactant. The other panels display the consumption of A by the surrounding B, at the times indicated. Dashed lines: B concentrations; dotted lines: A concentrations; continuous lines: C concentrations.

Another interesting configuration is the one in which a filament or lamella of width s_0 containing reactant A is compressed by the flow while sandwiched between two B regions (Fig. 5.2) in a straining flow. The two reaction fronts will advance until consuming all the A. Again the problem is described by (5.10), with $\phi(x, t = 0) = 1$ if $x \in (-s_0/2, s_0/2)$, and $\phi(x, t = 0) = -1$ elsewhere. The boundary conditions are $\phi(x \pm \infty, t) = -1$. The solution to the diffusion equation in the transformed variables τ and ξ can be written from expression (2.96). Initially the solution resembles two interfaces of the type (5.11), but at long times the form

$$\tilde{\phi}(\xi, \tau) \approx -1 + \frac{2s_0}{\sqrt{4\pi\tau}} e^{-\frac{\xi^2}{4\tau}} = -1 + \sqrt{\frac{2\lambda}{\pi D}} s_0 e^{-\lambda t} e^{-\frac{x^2 \lambda}{2D}} = \phi(x, t),$$

$$(5.19)$$

similar to (2.89) , is approached. The concentration distribution is obtained from (5.8)-(5.9) with $T = 1$. This consumption of the A striation is displayed in Fig. 5.2.

5.3　The fast binary reaction in complex flows

All the previous calculations referred to simple lamellar situations involving a single or a couple of interfaces. In more realistic flows, the interfaces will become stretched and folded, forming densely packed striations that sooner or later overlap and interact. Several studies of these situations are available. Jimenez and Martel (1991) considered the $A + B \rightarrow C$ reaction, in the fast reaction limit, in a temporally developing shear layer obtained from direct two-dimensional Navier-Stokes simulations (up to Re $= 1600$). Fluid containing pure A is initially located above the line $y = 0$, and the fluid containing B is below. Horizontal shear is applied and the fluid develops the Kelvin-Helmholtz instability, so that the interface is rolled up by the Kelvin-Helmholtz vortices. Diffusion-limited reaction occurs at this interface of increasingly complex geometry. The main observations were that the product distribution was located mainly on the neighborhood of the material interface obtained by advection of the initial flat interface, and that at each time the total amount of product in the system, M, scaled as $M \sim \text{Pe}^{-\alpha}$, with an exponent α that was in the range $0.4 - 0.5$ at initial times and that changed to $\alpha \approx 0.17$ at later times coinciding with the pairing of vortices in the flow, which introduced significant strain into the system.

Jimenez and Martel (1991) explained their observations by realizing that the developing material interface reaches a fractal shape during its evolution, and that the product distribution covers it with a strip of finite width w. This concept has been developed again in a variety of situations (Károlyi et al., 1999; Wonhas and Vassilicos, 2002; Tél et al., 2005). For the present binary reaction the calculations in the previous section, performed for isolated interfaces, can be used to estimate this width w. Depending on the type of flow it can have different time dependencies, but at a fixed time it always scales as $w \sim \mathrm{Pe}^{-1/2}$. From the definition of fractal dimension, a material line folded to form a fractal of dimension D_F, needs a number $N(w) \sim w^{-D_F}$ of elements of small diameter w to be covered. The length of the fractal scales thus as $\mathcal{L} = wN(w) \sim w^{1-D_F}$ when measured at resolution w, and the measured area scales as $\mathcal{A} = w^2 N(w) \sim w^{2-D_F}$. When w is the width of product distribution, the area occupied by such elements gives the amount of product in the system, which then scales as

$$ M \sim w^{2-D_F} \sim \mathrm{Pe}^{-(1-D_F/2)} \sim D^{1-D_F/2} \ . \qquad (5.20) $$

Then, $\alpha = 1 - D_F/2$. The analogous result in a $d-$dimensional flow would be $M \sim w^{d-D_F} \sim \mathrm{Pe}^{-(d-D_F)/2}$. Jimenez and Martel (1991) checked that under time evolution, the material line that initially separated the two reactants becomes a convoluted spiral-type object with an empirical value of D_F of about $1.6 \approx 5/3$, in agreement with the observed scaling of the product distribution. The authors explained these numbers by noticing that initially the interface (a line of $D_F = 1$) was rolled up by the Kelvin-Helmholtz vortices into sets of spirals similar to the one displayed in Fig. 2.19, to which a value of $D_F = 4/3$ can be assigned (Vassilicos and Hunt, 1991) and that further strain acting on the spirals, as happening during the pairing instabilities experienced by the flow, changes this value to $D_F = 5/3$.

It should be remarked here that the length of this finite-width reactive interface in the mixing layer is continuously increasing, so that, in a closed container, it will eventually fill the whole system and at long times one should have $D_F = 2$. But during the times analyzed in the numerical experiment, the interface occupied only a small proportion of the domain area, and the fractal scaling was robust. We also stress that it is not necessary for the fractal scaling to persist to arbitrarily small widths. It is enough to have an effective fractal structure down to a certain finite resolution fixed by the product distribution width, w.

A related situation was described by Wonhas and Vassilicos (2002). They considered the chlorine monoxide deactivation process:

$$ClO + NO_2 \rightarrow ClONO_2,$$

that is important in stratospheric ozone chemistry, in the presence of advection by observed atmospheric flows. The time evolution of the material line initially separating polar air (rich in ClO) from mid-latitude air (rich in NO_2) was also followed, and characterized by a fractal dimension D_F at each instant of time. D_F was seen to increase from $D_F = 1$ at short times towards $D_F = 2$ at long times, when it fills the space completely. At each fixed time, relationship (5.20) between the Péclet number or the diffusion coefficient and the amount M of product $ClONO_2$ was observed.

Within the reactive interface description, expression (5.18) obtained within the lamellar model in a straining flow can be used in the case of complex interfaces simply by replacing L by the interface length $\mathcal{L}$. The production $\mathcal{P}$, giving the effective rate of reaction, becomes

$$\mathcal{P} = \mathcal{L}p \sim \mathcal{L}\sqrt{\lambda D} \sim D^{\frac{1-D_F}{2}} D^{\frac{1}{2}} \sim \mathrm{Pe}^{-\left(1-\frac{D_F}{2}\right)}. \tag{5.21}$$

We note that the scaling with Pe also applies to shear-type flows.

Equation (5.21) is in fact equivalent to an interesting result which is valid for any reaction occurring on a line with a production which remains proportional to the length of that line. This is the case of the fast binary reaction discussed here, but also of other types such as autocatalytic, competitive, predator-prey, and others to be discussed in the following chapters. When the line becomes folded to form a fractal of dimension $D_F \in (1,2)$ (at least above a resolution of the order of the product width), we can write (Toroczkai et al., 1998; Károlyi et al., 1999; Tél et al., 2005):

$$\mathcal{P} = \mathcal{L}p \sim w^{1-D_F} \sim cM^{-\beta} \; , \; \beta = \frac{D_F - 1}{2 - D_F}, \tag{5.22}$$

where c involves geometric factors, and D and λ. We have used that the total amount of product M present in the system scales as $M \sim w^{2-D_F}$. The result $\mathcal{P} \approx cM^{-\beta}$ is also obtained for reactions occurring at interfaces of local dimension $d - 1$ in d-dimensional systems, with $\beta = (D_F - d + 1)/(d - D_F)$. When $D_F = 1$, so that the interface is a simple non-fractal line of finite length L, $\beta = 0$ and the production becomes $\mathcal{P} = gL$, independent of the width w.

The consequences of equations such as (5.22) or (5.21) (which state that production is proportional to interface length) have been discussed in detail by Toroczkai et al. (1998), Károlyi et al. (1999), or Tél et al. (2005) (with emphasis on the autocatalytic case). For example, one of the puzzling observations reported in references such as Edouard et al. (1996) or Mahadevan and Archer (2000) is that the observed reactivity of the simulated atmospheric chemistry or marine biological dynamics, respectively, increases when increasing the resolution of the computer model, i.e. when resolving smaller features. Since an increase in resolution is associated to a reduction of the numerical or effective diffusion, the dependence of the production term as $D^{(1-D_F)/2}$, which diverges as $D \to 0$, may provide an explanation for that: if the reaction occurs everywhere on a fractal support down to a minimum scale fixed by the resolution, resolving more details of the fractal interface will lead to more production. We stress however that the amount of product covering the fractal gets smaller when resolving better the filaments by reducing their width.

The situation is somehow similar to the effective diffusion: the presence of many spatial scales below the observational one increases the effective diffusion well above molecular values (Sect. 2.2.2) and may even change the character of diffusion turning it into anomalous (Sect. 2.8.1). Here, the convoluted geometry generated by the flow changes the production term in the chemical kinetics from what is expected in a fast binary reaction occurring at a smooth interface ($\beta = 0$) to an anomalous kinetics ($\beta > 0$).

We stress now the limitations of the approach presented in this section. First, we are describing with fractal concepts some objects, the reacting interfaces, which only have a finite range of scaling. As discussed above, this is not a problem as long as the scaling extends above the width of the product distribution. Recent work has formalized effective dimensions in this context (Tél et al., 2005; Károlyi and Tél, 2005; Károlyi and Tél, 2007). Second, fractal dimensions and widths may be time dependent. The example of the shear flow gives the situation in which the strip of product diffusively widens around the reacting interface, but irregular fluctuations will be always present in complex flows. This poses no problems on the results which describe scaling at a fixed time, but needs to be taken into account to describe time-dependence of production or of the amount of product. The theory can be nicely developed when the interfaces form a fractal set with well defined dimension which remains constant in time (Tél et al., 2005). This is what happens when the reaction in a chaotic open flow reaches its asymptotic state. In this case, an additional escape of fluid from the system at rate κ is present. This is analogous to the λ term in (5.16) or (5.17), which coincides there with the local strain rate or Lyapunov exponent, but which in an open chaotic flow will be smaller than λ. Combining escape with production, we can write for the amount of product in the system (Tél et al., 2005):

$$\dot{M} = -\kappa M + cM^{-\beta} \; . \tag{5.23}$$

As a third limitation, the results in this section are obtained in the limit of infinitely fast reaction. Deviations from this limit would need to be taken into account when addressing specific reaction schemes. Fourth, we have assumed that the interfaces, even if convoluted, are well separated, so that we have essentially $A = 1$ and $B = 1$ except in the narrow product width. When the interfaces become close enough they annihilate, as shown in the example of Fig. 5.2, and this interacting stage is not well described by the arguments in this section. They do not cover neither the filament formation and elongation process.

Within this interacting-filament regime of the fast binary reaction under mixing by a complex flow, the approach of the product amount to its final equilibrium value is expected to proceed exponentially. This can be argued by considering the equations for product width and interface effective dimension (Károlyi and Tél, 2005) or by noting from Eqs. (5.8)-(5.9) that the decay of the product concentration C towards its final value $T/2$ is described by the dynamics of the auxiliary field $\phi(x,t)$ which is the solution of the advection-diffusion equation (5.6). As discussed in Sect. 2.7.2 the decay of $\phi(x,t)$ at long times under chaotic mixing is typically exponential. The experimental study by Arratia and Gollub (2006) of a binary acid-base neutralization reaction was generally consistent with an exponential approach to steady state, but a faster behavior was observed at very long times. A more careful consideration of this late time regime would then be needed. Crossovers between several mechanisms (Sokolov and Blumen, 1991) as well as the effect of boundaries (Chertkov and Lebedev, 2003) would have to be probably included.

Chapter 6

Decay-type and Stable Reaction Dynamics in Flows

In the case of the fast binary reaction we could eliminate the reaction term from the reaction-diffusion-advection equation. But in general this is not possible. In this chapter we consider another class of chemical and biological activity for which some explicit analysis is still feasible. We consider the case in which the local-reaction dynamics has a unique stable steady state at every point in space. If this steady state concentration was the same everywhere, then it would be a trivial spatially uniform solution of the full reaction-diffusion-advection problem. However, when the local chemical equilibrium is not uniform in space, due to an imposed external inhomogeneity, the competition between the chemical and transport dynamics may lead to a complex spatial structure of the concentration field. As we will see in this chapter, for this class of chemical or biological systems the dominant processes that determine the main characteristics of the solutions are the advection and the reaction dynamics, while diffusion does not play a major role in the large Péclet number limit considered here. Thus diffusion can be neglected in a first approximation.

We first consider the impact of transport on global properties –

namely the total amount of the substances involved – of the chemical or biological dynamics. Next, we describe the spatial structures generated under turbulent or chaotic flows.

6.1 Stable reaction dynamics and its global steady state

We consider the problem

$$\frac{\partial C}{\partial t} + \mathbf{v} \cdot \nabla C = k_0 F(C; \mathbf{x}) + D\nabla^2 C \ , \tag{6.1}$$

where k_0 is a characteristic reaction rate. By selecting a characteristic flow spatial scale L and velocity U, we can render Eq. (6.1) dimensionless by measuring space in units of L and time in units of L/U:

$$\frac{\partial C}{\partial t} + \mathbf{v} \cdot \nabla C = \mathrm{Da} F(C; \mathbf{x}) + \frac{1}{\mathrm{Pe}} \nabla^2 C \ , \tag{6.2}$$

where $\mathrm{Da} \equiv k_0 L/U$ is the Damköhler number (which is the dimensionless ratio of the characteristic reaction time, $1/k_0$, and the advection time L/U). and $\mathrm{Pe} \equiv LU/D$ is the Péclet number. Assume that the reaction dynamics

$$\frac{dC}{dt} = \mathrm{Da} F(C; \mathbf{x}) \tag{6.3}$$

has a unique and stable local steady state $C^*(\mathbf{x})$ (i.e. $C^*(\mathbf{x})$ is the solution of $F(C^*; \mathbf{x}) = 0$ for fixed $\mathbf{x}$, and $F'[C^*(\mathbf{x})] < 0$ for any $\mathbf{x}$ in the domain considered). In the case of a multi-component system the corresponding conditions are that the set of equations

$$F_i(C_1, \ ... \ , C_N; \mathbf{x}) = 0, \quad i = 1, \ ... \ , N \tag{6.4}$$

has a single stable solution $\mathbf{C}^* \equiv \{C_i^*(\mathbf{x})\}$, and all the eigenvalues of the Jacobian matrix $J \equiv [\partial F_i/\partial C_j]$ evaluated at the steady state $\mathbf{C}^*$ are negative.

We will refer to this type of systems as decay-type reaction systems since the stability of the local equilibrium implies that deviations from the stable steady state, in the absence of transport processes, decay in time. When the concentration field is advected by the flow, there is a competition between the reaction dynamics that drives the system towards its local equilibrium, and the advection that maintains the non-equilibrium state by moving the fluid parcels around in a spatially non-uniform environment.

The simplest example of this type of reaction is the linear decay of a contaminant which is injected into the flow by a non-uniform spatially distributed source:

$$F(C; \mathbf{x}) = S(\mathbf{x}) - bC, \tag{6.5}$$

where $1/b$ is the average lifetime of the contaminant. This may represent spontaneous decay of an unstable chemical compound, a photochemical reaction, or a binary reaction with a substance present in a high excess concentration that remains approximately constant (see Sect. 3.1.2). The reaction kinetics (6.5) can be generalized to the nonlinear case with concentration-dependent decay rate, $b(C)$, for example, when an annihilation reaction $A + A \to \emptyset$ is considered and a second-order kinetics is appropriate. Chemical decay in flows is relevant for several applications, as for example the distribution of atmospheric pollutants or ozone decay in the stratosphere. Another problem, formally equivalent to (6.5), is the relaxation to a spatially non-uniform distribution, e.g. of temperature, according to

$$F(T; \mathbf{x}) = k_0[T_0(\mathbf{x}) - T]. \tag{6.6}$$

This type of dynamics is used, for example, in models of oceanic surface temperature distribution (Abraham and Bowen, 2002), with $T_0(\mathbf{x})$ representing an imposed atmospheric temperature field.

In the case of biological populations, a simple example in this category is the logistic population dynamics (Sect. 3.2.1) of oceanic plankton in an environment with spatially non-uniform carrying capacity (e.g. due to non-uniform light, temperature or nutrient availability)

$$F(P; \mathbf{x}) = P\left[1 - \frac{P}{K(\mathbf{x})}\right]. \tag{6.7}$$

The main characteristic of the chemical or biological field in a bounded domain is the average concentration in the stationary state.

Integrating (6.2) over the whole domain the contributions of the advective and diffusive transport terms vanish for typical boundary conditions (e.g. periodic or no flux) and

$$\frac{d}{dt}\int C(\mathbf{x})\mathbf{dx} = \int F[C(\mathbf{x}); \mathbf{x}]\mathbf{dx}, \qquad (6.8)$$

where the left hand side is zero in the steady state. When $F(C, \mathbf{x})$ is a linear function of C, as in (6.5), it follows that the total or the average concentration is independent of the flow and it is fully determined by the average source strength, $\langle C \rangle = \langle S \rangle / b$.

For general nonlinear reaction kinetics, (6.8) cannot be reduced to a closed equation for the average concentration and therefore $\langle C \rangle$ depends on the spatial structure of the field created by the flow. Two limiting cases can be considered. When the flow is very slow compared to the reaction, i.e. Da $\gg 1$, the chemical dynamics almost reaches its local equilibrium everywhere in space and therefore the average concentration is approximately equal to the average of the local equilibrium, $\langle C \rangle \approx \langle C^*(\mathbf{x}) \rangle$. In the opposite case of very fast mixing, Da $\ll 1$, the strong homogenizing effect of fast advection and diffusion produces a concentration field that is almost uniform in space. In this case, an approximation for the average concentration can be obtained by solving the averaged equation (6.8) for a constant concentration independent of $\mathbf{x}$

$$\int F(\langle C \rangle, \mathbf{x})d\mathbf{x} = 0. \qquad (6.9)$$

In the case of linear decay these two limits coincide, but in general they are different and for intermediate mixing rates the average concentration is a non-trivial function of the Damköhler number.

For example, in the case of an annihilation reaction with quadratic decay, i.e. $F(C) \equiv S(\mathbf{x}) - bC^2$, the slow stirring limit gives

$$\langle C \rangle_{\mathrm{Da} \gg 1} \simeq \langle C^*(\mathbf{x}) \rangle = b^{-1/2} \left\langle \sqrt{S(\mathbf{x})} \right\rangle, \qquad (6.10)$$

while in the fast stirring case we have to solve $\int dx\, (S(x) - b\langle C \rangle^2) = 0$, which gives

$$\langle C \rangle_{\mathrm{Da} \ll 1} \simeq b^{-1/2} \sqrt{\langle S(\mathbf{x}) \rangle}. \qquad (6.11)$$

By Jensen's inequality it is easy to show that $\langle C \rangle_{\mathrm{Da}\gg1} \le \langle C \rangle_{\mathrm{Da}\ll1}$. This reflects the lower frequency of annihilation encounters and slower overall decay when the reactants are uniformly distributed, compared to the case of a distribution concentrated in some small regions. Therefore, the steady state average concentration, or total amount, is an increasing function of the stirring rate in this nonlinear annihilation process. Such change of the overall reactivity of a second order reaction due to chaotic mixing was observed experimentally by Paireau and Tabeling (1997).

We can apply the same analysis to the logistic plankton population dynamics with non-uniform carrying capacity (Eq. (6.7)) that was studied by McKiver and Neufeld (2009). In the slow mixing limit the equilibrium distribution is close to the local carrying capacity, and the average plankton concentration is $\langle P \rangle \approx \langle K(\mathbf{x}) \rangle$. When mixing is much faster than the plankton growth rate, the plankton density is almost uniform in space so that $\langle P^2 \rangle \simeq \langle P \rangle^2$, and averaging gives

$$\frac{d\langle P \rangle}{dt} \approx \mathrm{Da} \left[\langle P \rangle - \langle P \rangle^2 \left\langle \frac{1}{K(\mathbf{x})} \right\rangle \right]. \tag{6.12}$$

From here the steady state average plankton concentration is obtained as the harmonic mean of the carrying capacity, $\langle P \rangle \approx 1/\langle K^{-1} \rangle$. Again using Jensen's inequality it can be shown that this is always smaller than the mean carrying capacity. Thus, interpolating the two limiting cases we can expect that in a system with non-uniform carrying capacity the average plankton concentration decreases with the mixing rate, i.e. increases with Da. This is confirmed by numerical simulations as shown in Fig. 6.1. Similar behavior was observed in a plankton model with spatially non-uniform growth rate by Birch et al. (2007). Note that the monotonously decreasing form of the average concentration as a function of stirring rate is not universally valid for all kinds of nonlinear chemical or biological dynamics. When the interactions between different species is taken into account (e.g. in PZ or NPZ plankton models), mixing may have different effects depending on the particular functional form of the local population dynamics. For an example where the plankton concentration increases with stirring see e.g. Pasquero et al. (2005). We stress that the above averaging procedure and analysis can be in principle applied to any type of reaction dynamics, not just stable or decaying ones. But the fact that non-decaying dynamics necessarily implies the presence of several chemical species makes the formal results less explicit.

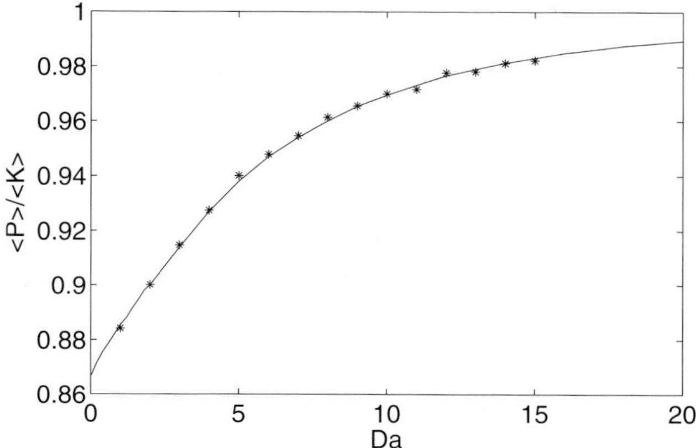

Figure 6.1: Average plankton population density as a function of the Damköhler number for logistic growth with non-uniform carrying capacity of the form $K(x,y) = K_0 + \delta \sin(2\pi x)\sin(2\pi y)$ and chaotic mixing in the time-periodic sine-flow of Eq. (2.66). The continuous line represents results from the solution of the full partial differential equation with diffusion (Pe $\simeq 10^4$) and stars (*) show the time-averaged plankton populations calculated from the non-diffusive Lagrangian representation.

It was first pointed out in the context of plankton population dynamics (Abraham, 1998) that the spatial structure in such systems is primarily due to the interplay between advection and the local population dynamics (or reaction kinetics in the case of chemical systems) and therefore the spatial distribution can be more easily analyzed by considering the non-diffusive limit. More precisely, in reaction-diffusion-advection systems with stable or decay type reaction dynamics the Pe $\rightarrow \infty$ limit is not singular. Therefore the weak diffusion term can be neglected and the solution of the simpler reaction-advection problem provides a good approximation to the concentration field. This approach does not hold in general, since for other types of reaction dynamics, or even in the case of a non-reactive passive scalar field, diffusion is a singular perturbation. In the passive scalar problem the concentration fluctuations created by a large scale source are transferred to smaller scales by chaotic advection, and in the absence of diffusion there is no mechanism to dissipate these fluctuations, therefore without diffusion a forced dissipative equilibrium cannot be established. In the case of a chemical or biological field with a decay-type dynamics the concentration fluctuations decay at all scales and eventually become negligible when they reach very small scales. Therefore the reaction dynamics and advection are sufficient to reach a statistical equilibrium. This is not altered significantly by weak diffusion, that in the case of large Péclet number only affects the smallest scales in the concentration field.

This simplification allows us to use a convenient Lagrangian representation of the problem (6.2). Without the diffusion term the solution can be expressed by the method of characteristics through a set of ordinary differential equations

$$\frac{d\mathbf{r}}{dt} = \mathbf{v}(\mathbf{r}, t), \quad \frac{dC}{dt} = \mathrm{Da}F(C, \mathbf{r}), \qquad (6.13)$$

were, as in Sect. 2.1.1, $C(t) \equiv C(\mathbf{x} = \mathbf{r}(t), t)$ is the concentration carried by a fluid parcel moving with the flow along the trajectory $\mathbf{r}(t)$. In an ergodic flow where a single trajectory uniformly covers the whole domain as $t \to \infty$ the spatial average of the concentration field can be obtained as a time-averaged concentration corresponding to a moving fluid parcel. This is confirmed numerically in Fig. 6.1 that shows the solutions of the full reaction-advection-diffusion system for logistic population dynamics with a very small diffusion coefficient, and almost identical results obtained from the corresponding Lagrangian representation computed from a single fluid element path.

The dependence of the average concentration on the Damköhler number can also be interpreted within the Lagrangian formulation. For example, the logistic growth function of the plankton population dynamics (Eq. (6.7)) is concave near the steady state $P = K$, i.e. the plankton population reacts more quickly when the carrying capacity is below the actual plankton density, than in the opposite case when higher carrying capacity allows for increase of the plankton concentration. Due to this asymmetric nonlinear response the net effect of the fluctuations around the local equilibrium is a lower average concentration than the average carrying capacity. The effect of fluctuations is more pronounced when the stirring is fast, since the plankton concentration within the fluid parcels has less time to reach the local carrying capacity.

6.2   The spectrum of decaying scalar in a flow

After discussing the effect of the flow on the average concentration, we turn to the analysis of the statistics of the concentration fluctuations in such decay-type reaction systems. Even in the case of linear decay, where the average concentration is not affected by mixing, the fluctuations depend on the advecting flow. The distribution of the fluctuations over different length scales can be characterized by the power spectral density of the concentration field (taken along a one-dimensional section $C(x)$, for simplicity) defined as

$$\Gamma(k) = \left| \frac{1}{\sqrt{2\pi}} \int C(x)e^{-ikx}dx \right|^2 = \frac{1}{2\pi}\hat{C}(k)\hat{C}^*(k), \qquad (6.14)$$

where $\hat{C}(k)$ is the Fourier transform of $C(x)$.

The distribution of a decaying scalar field advected by a turbulent flow was studied by Corrsin (1961) who generalized the Obukhov-Corrsin theory of passive scalar turbulence for the linear decay problem $F(C) = S(\mathbf{x}) - bC$. As in the case of the passive non-decaying scalar field, depending on the length scales considered, one can identify inertial-convective and viscous-convective regimes with qualitatively different characteristics.

6.2.1   The inertial-convective range

The inertial-convective range of three-dimensional turbulence covers the range of length scales where inertial forces dominate over viscous forces in the dynamics of the velocity field and advection is the dominant transport process with respect to diffusion. This is valid below the integral scale and limited at small scales by the larger of the Kolmogorov scale (η) or the diffusive scale (l_D). For

Schmidt numbers larger than one, $Sc = \nu/D > 1$, the diffusive scale is smaller than the Kolmogorov scale and therefore the inertial-convective range coincides with the entire inertial range of the turbulent flow. In this regime the characteristic time of the velocity field corresponding to a given lengthscale (l) decreases towards smaller scales as $\tau(l) \simeq l/\delta v(l) \simeq l^{2/3}\epsilon^{-1/3}$, or as a function of the wavenumber $\tau(k) \simeq \epsilon^{-1/3}k^{-2/3}$, where ϵ is the energy dissipation rate.

The concentration fluctuations produced by the source at large scales, which is assumed to have variations on lengthscales comparable to the integral scale of the flow, are distorted by advection generating smaller scales in the concentration field. Therefore the characteristic wavenumber associated to a wavepacket representing concentration fluctuations in a narrow spectral range increases in time as

$$\frac{dk}{dt} \approx \frac{k}{\tau(k)} = \epsilon^{1/3}k^{5/3}. \tag{6.15}$$

This is the same as for a non-reactive passive scalar since the linear decay affects all scales equally, so it does not change the shape of the concentration iso-contours but only reduces the contrast between them.

The power spectrum evolves as

$$\frac{\partial \Gamma(k)}{\partial t} = \hat{C}^* \frac{\partial \hat{C}}{\partial t} + C \frac{\partial \hat{C}^*}{\partial t}, \tag{6.16}$$

and taking the Fourier transform of the advection-diffusion equation for the decaying scalar gives

$$\frac{\partial \hat{C}}{\partial t} + \mathcal{F}[\mathbf{v} \cdot \nabla C] = \hat{S} - b\hat{C} - k^2 D\hat{C}, \tag{6.17}$$

where $\mathcal{F}$ is the Fourier transform operator and $\hat{S} = \mathcal{F}[\mathcal{S}]$ is assumed to contain only low wavenumbers. When we substitute (6.17) in the equation for the power spectrum (6.16) the terms on the right, representing decay and diffusion, are local in wavenumber space, while the advection term is responsible for the transfer of fluctuations from low to high wavenumbers. A phenomenological representation of this contribution of advection to the power spectrum can be obtaining by considering it as a kind of advection in k-space, at a speed given by

(6.15). Thus the advective flux of the power in k space is $\Gamma(k)dk/dt$, and the continuity equation for the power of fluctuations (compare with Eq. (1.1)) will read

$$\frac{\partial \Gamma(k)}{\partial t} + \frac{\partial}{\partial k}\left[\frac{dk}{dt}\Gamma(k)\right] = -2b\Gamma(k) - 2Dk^2\Gamma(k). \qquad (6.18)$$

The source term can be included as a constant flux boundary condition at the lowest wavenumber, $k_S = 2\pi/L_S$. The terms on the right represent the decay of fluctuations due to (i) the reaction acting uniformly on all scales, and (ii) due to diffusion, that becomes stronger at large wavenumbers. For large Pe the diffusive decay is relatively weak in the inertial convective range and can be neglected for simplicity.

In a statistically stationary state $\partial\Gamma/\partial t = 0$. Substituting (6.15) into (6.18) and integrating over k gives the solution

$$\Gamma(k) = Ak^{-5/3}\exp\left(\frac{3b}{\epsilon^{1/3}k^{2/3}}\right), \qquad (6.19)$$

where the constant A depends on the source strength. The exponential factor defines a characteristic wavenumber $k_r = b^{3/2}\epsilon^{-1/2}$, associated to a reaction length scale $l_r = 2\pi/k_r$. At scales smaller than l_r, i.e. in the range of high wavenumbers $k_r \ll k \ll 2\pi/\eta$, the flow is much faster than the chemical decay, therefore the power spectrum is approximately $k^{-5/3}$, indicating that the concentration field behaves like a passive scalar. At larger scales, or low wavenumbers $k < k_r$, the spectrum falls off steeply as fluctuations are damped by the chemical decay. The reaction length scale l_r decreases as the decay rate b is increased and the passive scalar-like $k^{-5/3}$ regime may dissapear completely when l_r becomes smaller than the Kolmogorov scale. The shape of the power spectrum is shown in Fig. 6.2.

6.2.2   The viscous-convective range

For large Schmidt numbers the diffusion is weaker than the viscosity, therefore below the Kolmogorov scale there is a viscous-convective range in which the transport is dominated by chaotic advection in a

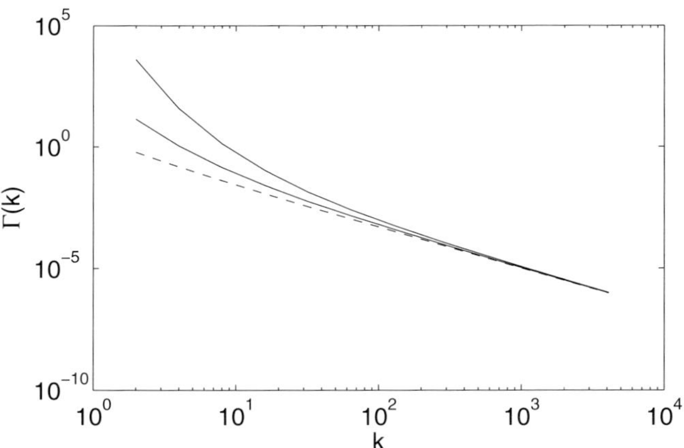

Figure 6.2: Sketch of the spectrum (6.19) for two different values of the decay rate (b is larger (smaller) for the top (bottom) curve). Dashed line indicates the $k^{-5/3}$ Obukhov spectrum of a non-reactive passive scalar for comparison.

smooth flow. Thus the characteristic wavenumber of a wavepacket of concentration fluctuation increases exponentially as

$$k(t) = k_0 e^{\lambda t}, \quad \text{or} \quad \frac{dk}{dt} = k\lambda, \tag{6.20}$$

where λ is the Lyapunov exponent of the chaotic advection and can be estimated as the inverse timescale of the eddies corresponding in size to the Kolmogorov scale, $\lambda \simeq 1/\tau_\eta = (\nu/\epsilon)^{1/2}$. Note that in contrast with the scale-dependent eddy timescale of the inertial range, in this viscosity-dominated regime the characteristic timescale of the flow, λ^{-1}, is independent of the wavenumber. Substituting (6.20) into Eq. (6.18) the steady state solution is of the form

$$\Gamma(k) \approx A k^{-1-2b/\lambda} \exp\left(\frac{Dk^2}{\lambda}\right). \tag{6.21}$$

Here we neglected intermittency corrections which are indeed present due to the distribution of Lyapunov exponents. This will be discussed later. In the special case $b = 0$ we recover the Batchelor k^{-1} spectrum of a passive scalar with a cut-off at the diffusive scale $l_D = \sqrt{D/\lambda}$.

Since $b > 0$ the power spectrum is steeper than k^{-1} as a consequence of the chemical decay, as has been verified numerically by Nam et al. (1999). Just like the Batchelor spectrum, the spectral form (6.21) is expected to hold for any smooth velocity field that generates chaotic advection. Thus, it remains valid even in the absence of an inertial range, as is the case for unsteady laminar flows or in the enstrophy cascade range of two-dimensional turbulence, which is relevant for geophysical flows.

6.3 Smooth and filamental distributions

The chemical decay $F(C) = S(\mathbf{x}) - bC$ has also a clearly visible effect on the structure of the chemical field in the real physical space (Neufeld et al., 1999, 2000). To show this we consider the dependence of the concentration differences between two points on the distance separating them

$$\delta C(l) \equiv |C(\mathbf{x} + \mathbf{l}) - C(\mathbf{x})| \sim |\mathbf{l}|^{\alpha} \tag{6.22}$$

where the distance l is assumed to be small compared to the characteristic length scale of the flow. The exponent $0 < \alpha < 1$ is the Hölder exponent, that characterizes the *roughness* of the concentration field. For a smooth distribution $\alpha = 1$, and the roughness increases as α becomes smaller. When $\alpha < 1$ the graph of the function $C(\mathbf{x})$ along a one-dimensional section is a self-affine fractal with dimension $D_f = 2 - \alpha$, $1 < D_f < 2$. A classic example of such continuous but nowhere differentiable function is the Weierstrass function defined as

$$W(x) = \sum_{n=0}^{\infty} a^n \cos(2\pi b^n x) \tag{6.23}$$

where $0 < a < 1 < b$ and $ab \geq 1$. The graph of this function has dimension (Hunt, 1998)

$$D_f = 2 + \frac{\log a}{\log b} , \quad \text{and} \quad \alpha = 2 - D_f. \tag{6.24}$$

Examples are shown in Fig. 6.3.

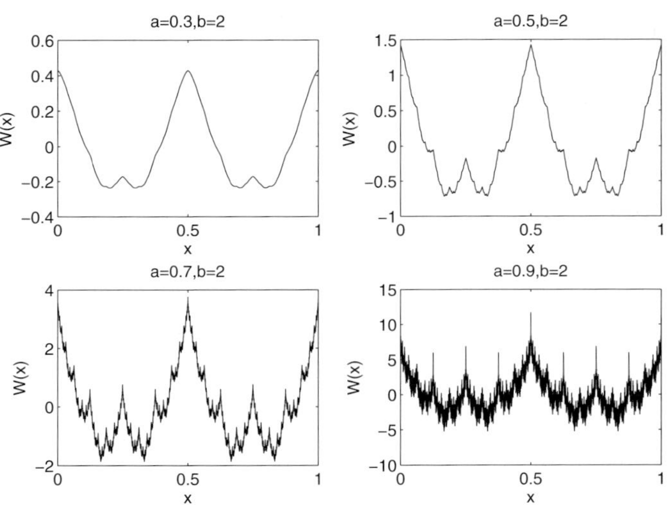

Figure 6.3: Weierstrass functions (6.23) with different degrees of roughness.

Neufeld et al. (1999) have shown that the roughness exponent α of the decaying chemical field is a function of the decay rate and of the Lyapunov exponent of the advection. In the large Péclet number limit we can neglect diffusion and set $D = 0$ so that the concentration field can be described by the Lagrangian representation (6.13) that follows the chemical dynamics within the fluid parcels advected on chaotic trajectories in the flow

$$\frac{d\mathbf{r}}{dt} = \mathbf{v}(\mathbf{r}, t), \quad \frac{dC}{dt} = S(\mathbf{r}) - bC. \tag{6.25}$$

To obtain the chemical concentration $C(\mathbf{x}, t)$ at a point $\mathbf{x}$ at time t one first needs to integrate the advection equation backwards in time to find the past trajectory of the fluid parcel, $\mathbf{r}(t - \tau)$ for $0 < \tau < t$ satisfying $\mathbf{r}(t) = \mathbf{x}$. Then the second equation in (6.25) for the chemical dynamics along this trajectory can be solved as (compare with (2.6))

$$C(\mathbf{x}, t) = C[\mathbf{r}(0), 0]e^{-bt} + \int_0^t S[\mathbf{r}(t - \tau)]e^{-b\tau}d\tau . \tag{6.26}$$

The contribution of the first term decays in time and can be neglected when the stationary state is considered, that is thus independent of the initial concentration field. The concentration difference between two points separated by a small distance $\mathbf{l}$ can be written as

$$\delta C \equiv C(\mathbf{x} + \mathbf{l}, t) - C(\mathbf{x}, t) = \int_0^t \delta S_{t-\tau} e^{-b\tau} d\tau \qquad (6.27)$$

where

$$\delta S_{t-\tau} \equiv S[\mathbf{r}(t - \tau) + \boldsymbol{\delta}(t - \tau)] - S[\mathbf{r}(t - \tau)] \qquad (6.28)$$

and $\boldsymbol{\delta}(t - \tau)$ is the time dependent separation vector connecting the two fluid elements defined for the time interval $0 < \tau < t$ and satisfying $\boldsymbol{\delta}(t) = \mathbf{l}$ at the final time. In a chaotic flow the separation between two fluid elements along the past trajectories increases exponentially backwards in time at an average rate given by the Lyapunov exponent, $|\boldsymbol{\delta}(t - \tau)| \sim l \exp(\lambda\tau)$, with $l = |\mathbf{l}|$. This exponential form is valid while the distance between the two fluid elements is smaller than the integral scale L, i.e. up to a saturation time $\tau_s = \lambda^{-1} \ln(L/l)$, after which $|\boldsymbol{\delta}(t - \tau)|$ saturates and fluctuates around L (in an unbounded flow $|\boldsymbol{\delta}|$ may increase further, but slower than exponentially, e.g. as a power law according to the Richardson law or to Taylor diffusion (Chapter 2). The exponential separation backwards in time is valid for almost all directions of the final separation vector $\mathbf{l}$, except for the special case when its direction coincides with the orientation of the unstable foliation at the selected point $\mathbf{x}$. We come back to this special case later.

The integral (6.27) can be separated into two parts. In the exponential separation regime, $\tau < \tau_s$, we have $|\boldsymbol{\delta}| \ll L$ and δS can be approximated by a Taylor expansion. For $\tau > \tau_s$ the two trajectories separated by a distance comparable to the characteristic lengthscale of the flow are uncorrelated and δS is a stationary chaotic function, thus

$$\delta C(l) \simeq l \int_0^{\tau_s} [\mathbf{n} \cdot \nabla S]_{t-\tau} e^{(\lambda - b)\tau} d\tau + \int_{\tau_s}^t \delta S_{t-\tau} e^{-b\tau} d\tau \qquad (6.29)$$

where $\mathbf{n}$ is a time-dependent unit vector pointing in the direction of $\boldsymbol{\delta}$. Changing the integration variable to $\delta \equiv |\boldsymbol{\delta}| = l \exp(\lambda t)$ in the first integral and introducing $\hat{\tau} \equiv \tau - \tau_s$ in the second term we get

$$\delta C(l) \simeq l^{b/\lambda}\frac{1}{\lambda}\int_l^L [\mathbf{n}\cdot\nabla S]_\delta \delta^{-b/\lambda}d\delta + \left(\frac{l}{L}\right)^{b/\lambda}\int_0^{t-\tau_s}\delta S_{\hat{\tau}}e^{-b\hat{\tau}}d\hat{\tau}$$

(6.30)

where $\mathbf{n}\cdot\nabla S$ is a bounded aperiodically fluctuating function of δ. In the stationary state when $t \gg \tau_s$ the upper limit of the second integral can be extended to infinity, i.e. by taking the limit $t \to \infty$ for fixed l, so that it becomes independent of l. In the first term there are two possibilities. When $b > \lambda$ the dominant contribution for small l comes from the lower limit $\delta \approx l$, so the integral is of order $l^{1-b/\lambda}$. In this case the first term dominates in the $l \to 0$ limit, $\delta C(l)$ is proportional to l and the Hölder exponent is $\alpha = 1$.

Thus, when the decay is faster than the advective separation rate the concentration field remains a smooth differentiable function in the Pe $\to \infty$ limit. In the other case, when $b < \lambda$, the dominant contribution to the first integral comes from the upper limit, which is independent of l and both terms in (6.30) scale as

$$|\delta C(l)| \sim l^{b/\lambda}.$$

(6.31)

This corresponds to a Hölder exponent $\alpha = b/\lambda$ smaller than one, indicating that the concentration field has a non-differentiable "rough" structure with cusp-like singularities everywhere. Summarizing, the Hölder exponent of the concentration field is given by

$$\alpha = \min\left\{\frac{b}{\lambda}, 1\right\},$$

(6.32)

indicating that a transition from smooth to rough structure takes place at $\lambda = b$.

In the special case when the direction of the final separation is along the unstable foliation of the chaotic advection the corresponding backward trajectories are convergent, $|\boldsymbol{\delta}(t-\tau)| \simeq l\exp(-\lambda\tau)$. Therefore we should replace λ with $-\lambda$ in (6.30) and, ignoring the second term since there is no saturation time, we obtain $|\delta C(l)| \sim l$. Thus, in the direction of the unstable foliation the concentration field is always smooth at every point in space. When $b < \lambda$ the existence of a smooth direction in an otherwise rough concentration field results in a characteristic filamental structure with a strong local anisotropy as shown in Fig. 6.4.

We note that the above results are not limited to the case of linear decay, but also apply to any kind of decay-type or stable reaction dynamics in a flow with chaotic advection (Chertkov, 1999; Hernández-García et al., 2002). In such systems where the reaction dynamics is nonlinear, the decay rate b should be replaced by the absolute value of the negative Lyapunov exponent of the Lagrangian chemical dynamics given by the second equation in (6.25), that represents the average decay rate of small perturbations in the chemical concentration along the trajectory of a fluid element.

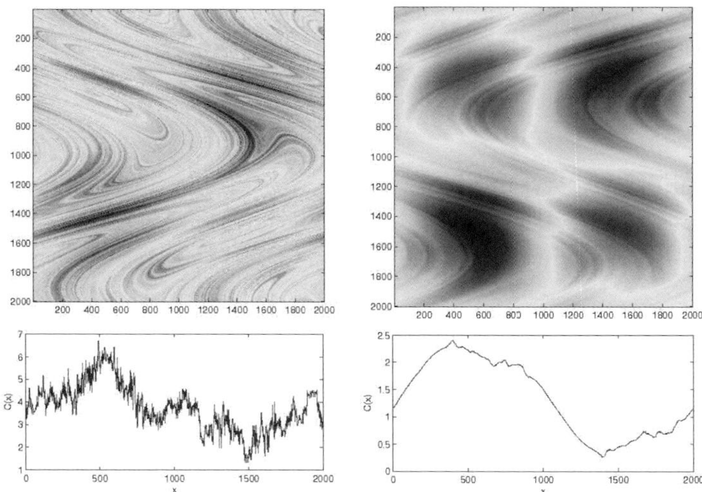

Figure 6.4: Filamental (left) and smooth (right) distributions of a decaying tracer in the chaotic sinusoidal shear flow with source distribution of the form $S(x, y) = \sin(2\pi x/L)\sin(2\pi y/L)$ for two different values of the decay rate. Lower panels show the fluctuations of the concentration along one-dimensional sections for both cases.

In a system with finite Péclet number the singular structure is smoothed out at small scales by diffusion. Thus, below the diffusive scale $l_D = \sqrt{D/\lambda}$ the concentration field is smooth, nevertheless the roughness persists and (6.32) remains valid in the range $l_D \ll l \ll L$ (López and Hernández-García, 2002).

Interestingly, the Weierstrass function defined by Eq. (6.23), originally considered as an example of an abstract "pathological" function (see Fig. 6.3), can be obtained as an exact solution of a simple discrete-time model of chemical decay and chaotic transport when advection is modelled by the so-called baker map (see e.g. Ott (1993)), which is a simple model of chaotic dynamics produced by stretching and folding. The baker transformation acts on the unit square by stretching it into a rectangle of length 2 and width 1/2, that is then cut into two pieces of unit length and placed back to cover the original square, as shown in Fig. 6.5. If we assume that the concentration field is initially uniform along the stretching direction, in this case the y axis, and is periodic along the contracting direction x, then the advection by the baker transformation is given by

$$C(x) \rightarrow C(2x \quad \mathrm{mod}\ 1). \tag{6.33}$$

The uniform stretching by a factor two in each iteration results in a Lyapunov exponent $\lambda = \ln 2$. The superimposed discrete time chemical decay with a source can be modelled as a reaction step of the form

$$C(x) \rightarrow S(x) + aC(x) \tag{6.34}$$

where $0 < a < 1$ and corresponds to a chemical decay rate $b = -\ln a$. Choosing the source function as $S(x) = \cos(2\pi x)$ the asymptotic solution coincides with the Weierstrass function in (6.23) with $b = 2$ and has Hölder exponent $\alpha = -\ln a / \ln 2$ (Eq. (6.24)).

6.4 Structure functions, multifractality and intermittency

The results in the previous section were based on the assumption that the separation rate of the fluid elements is well represented by a single Lyapunov exponent. However, the finite-time Lyapunov exponents (Sect. 2.5.1) may deviate from the asymptotic value and this can affect average properties of the concentration fluctuations

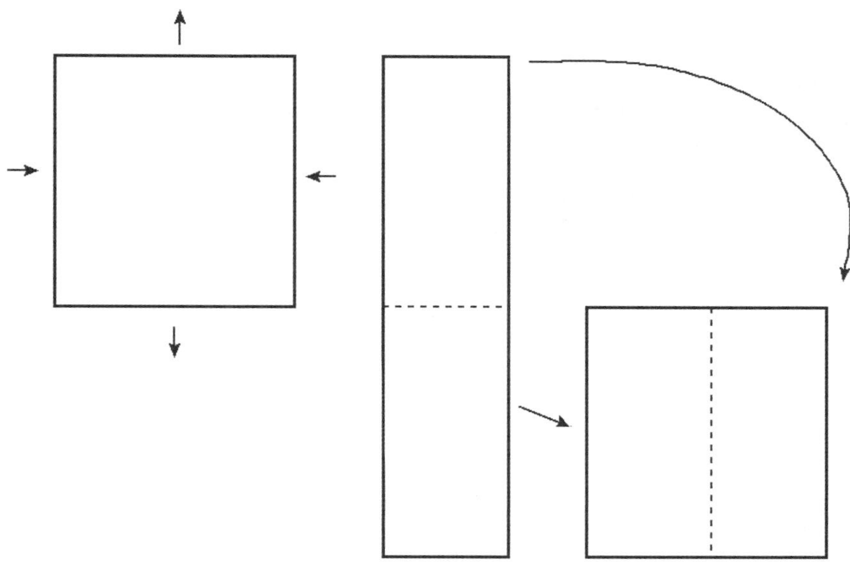

Figure 6.5: Sketch of the area preserving baker's map. The square is squeezed in the horizontal and stretched in the vertical direction, then it is cut in half to cover the original unit square.

(Chertkov, 1998; Neufeld et al., 2000). To characterize the intermittent fluctuations we can use the structure functions of order q defined as

$$S_q(l) \equiv \langle |C(\mathbf{x}+\mathbf{l}) - C(\mathbf{x})|^q \rangle \tag{6.35}$$

where q is a positive parameter and $\langle\ \rangle$ represents average in space over $\mathbf{x}$. The structure functions satisfy the scaling relation

$$S_q(l) \sim |\mathbf{l}|^{\zeta_q} \tag{6.36}$$

that defines the set of scaling exponents ζ_q.

If the concentration field had exactly the same local Hölder exponent (α) everywhere in space, the spatial averaging would be irrelevant and the scaling exponents would be given by $\zeta_q = q\alpha$. This is a valid approximation for small q, but typically there are corrections that lead to an anomalous scaling when the fluctuations of the finite-time Lyapunov exponents is taken into account. As time increases the finite-time Lyapunov exponents are more and more concentrated around the asymptotic value λ^∞, and their distribution follows the standard form (2.75)

$$P(\lambda, t) \sim t^{1/2} e^{-G(\lambda)t} \tag{6.37}$$

where $G(\lambda)$ satisfies $G(\lambda^\infty) = G'(\lambda^\infty) = 0$. We have shown in Sect. 2.5.1 that the fluid elements experiencing an anomalous Lyapunov exponent of value λ in the $t \to \infty$ limit are restricted to a fractal set of dimension given by (2.79):

$$D_f(\lambda) = 2 - \frac{G(\lambda)}{\lambda}. \tag{6.38}$$

This multiplicity of intertwined fractal sets with different properties is said to form a *multifractal*. Although the area covered by such sets is zero, they have a non-negligible contribution to the structure functions. To see this we note that the averaging in (6.35) can be expressed as an integral over the contributions coming from fluid elements characterized by different Lyapunov exponents. Dividing the domain into small non-overlapping regions of size l, the number of regions corresponding to a given value of λ which give rise to a local Hölder exponent $\alpha(\lambda)$ scales as $N(l) \sim l^{-D_f(\lambda)}$. In two dimensions the total number of regions of size l is $(L/l)^2$. Using $\delta C(l) \sim l^{\alpha(\lambda)}$ and substituting (6.32) for the local Hölder exponent, the structure function can be written as

$$S_q(l) \simeq \int_{\lambda_{min}}^{\lambda_{max}} l^{2-D_f(\lambda)} |l^{\alpha(\lambda)}|^q d\lambda$$

$$= \int_{\lambda_{min}}^{b} l^{2-D_f(\lambda)+q} d\lambda + \int_{b}^{\lambda_{max}} l^{2-D_f(\lambda)+qb/\lambda} d\lambda \tag{6.39}$$

In the limit $l \to 0$ the dominant contribution comes from a saddle point and the scaling exponents are

$$\begin{aligned}
\zeta_q &= \min_\lambda \left\{ 2 - D_f(\lambda) + q, \frac{qb}{\lambda} + 2 - D_f(\lambda) \right\} \\
&= \min_\lambda \left\{ q, \frac{qb}{\lambda} + 2 - D_f(\lambda) \right\} \\
&= \min_\lambda \left\{ q, \frac{qb + G(\lambda)}{\lambda} \right\}.
\end{aligned} \tag{6.40}$$

The minimum occurs at a λ_q that satisfies the equation

$$G'(\lambda_q) = \frac{qb + G(\lambda_q)}{\lambda_q}. \tag{6.41}$$

For small q, λ_q is close to the asymptotic Lyapunov exponent, but as q increases the dominant contribution comes from larger than average Lyapunov exponents $\lambda_q > \lambda^\infty$, representing rare but more singular local roughness. Varying the parameter q selects sets, within this multifractal hierarchy, with different Lyapunov (and Hölder) exponents, that are responsible for the dominant contribution to the structure function of order q. This results in a nonlinear dependence of the scaling exponents ζ_q on the order q, known as *anomalous scaling*. It is due to the so called *intermittency*, i.e. that low probability components, in this case very rare dispersion rates, contribute to averaged quantities like the structure functions because they produce very strong singularities at certain locations.

If q is relatively small then λ_q is close to the average Lyapunov exponent and the distribution of the finite-time Lyapunov exponents around λ^∞ can be approximated by a Gaussian form (2.76) which leads to

$$G(\lambda) \approx \frac{(\lambda - \lambda^\infty)^2}{2\Delta} \tag{6.42}$$

where the parameter Δ charaterizes the spread of the distribution. Substituting this into (6.41) and (6.40) the scaling exponents can be calculated explicitly as

$$\zeta_q = \sqrt{\frac{(\lambda^\infty)^2 + 2qb\Delta}{\Delta^2}} - \frac{\lambda^\infty}{\Delta}. \tag{6.43}$$

This result is exact in the limit of very short correlation time of the velocity field corresponding to the so called Kraichnan model (Chertkov, 1998).

Abraham and Bowen (2002) applied the above theoretical description to an analysis of ocean data from satellite observations (Fig. 6.6) and showed that the distribution of sea-surface temperature modelled as a relaxation to the atmospheric temperature is consistent with the anomalous scaling exponents given by (6.43).

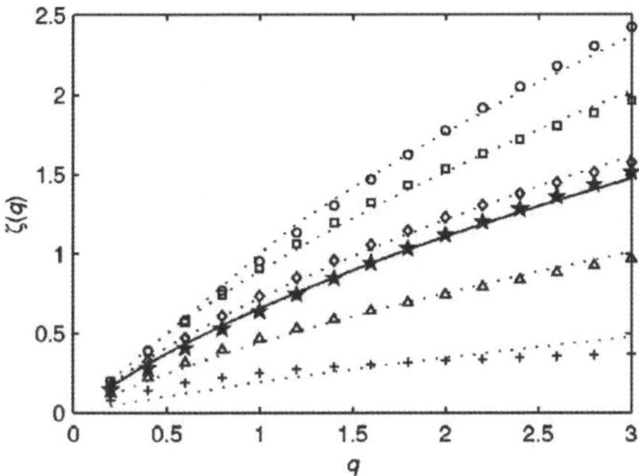

Figure 6.6: Anomalous scaling exponents of sea-surface temperature from Abraham and Bowen (2002). The solid stars are the exponents obtained from satellite data and the other symbols represent scaling exponents calculated from simulating the sea-surface temperature as a linear relaxation with different relaxation rates.

Thus, this type of relationships could be used to infer relaxation rates, or other quantitative characteristics of geophysical flows.

Nam et al. (1999) have shown that the intermittent inhomogeneity of the dispersion rate also leads to a correction for the exponent of the power spectrum (6.21). The power spectrum can be considered as a superposition of contributions from wavepackets of concentration fluctuations associated to trajectories that have experienced different history of dispersion rates distributed according to (6.37). The proportion of wavepackets with a cumulative dispersion rate λt after time t decays as $\exp(-G(\lambda)t)$. Thus, the contribution of wavepackets for which the finite-time Lyapunov exponent has a particular value λ satisfies Eq. (6.18) in a generalized form

$$\frac{\partial \Gamma_\lambda(k)}{\partial t} + \frac{\partial}{\partial k}\left[\lambda k \Gamma_\lambda(k)\right] = -2b\Gamma_\lambda(k) - 2Dk^2\Gamma_\lambda(k) - G(\lambda)\Gamma_\lambda(k) \tag{6.44}$$

where $k_\lambda(t) \simeq k_0 \exp(\lambda t)$. The steady state solution of (6.44) is

$$\Gamma_\lambda(k) = k^{-1-\frac{2b+G(\lambda)}{\lambda}} \exp\left(\frac{Dk^2}{\lambda}\right). \tag{6.45}$$

The dominant contribution to the full spectral density at large k comes from wavepackets whose spectrum decays most slowly in the $k \to \infty$ limit, hence

$$\Gamma(k) \sim k^{-\gamma} \quad \text{where} \quad \gamma = 1 + \min_{\lambda}\left\{\frac{2b + G(\lambda)}{\lambda}\right\}. \qquad (6.46)$$

When the dispersion rates are close to uniform and concentrated to a narrow band around λ^{∞} the Corrsin result $\gamma = 1 + 2b/\lambda^{\infty}$ is recovered. Otherwise the minimum in (6.46) is achieved at $\lambda > \lambda^{\infty}$, i.e. the intermittency correction leads to a somewhat less steep spectral slope. This has been verified by Nam et al. (1999) using numerical simulations of chemical decay in a smooth two-dimensional flow. Note that (6.46) can also be obtained directly from the exact relationship between the spectral slope and the second order scaling exponent $\gamma = 1 + \zeta_2$.

In the following we consider two special cases of decay type reactions that exhibit strong intermittency. In the first case this is due to localized mixing in open flows, while in our second example it is a result of very fast reaction restricted to certain "resetting" regions.

Open flows

Open flows with a localized mixing region of the type discussed in Sect. 2.6 provide an example of strongly non-uniform distribution of the Lyapunov exponents. In such flows, permanent chaotic advection is restricted to a fractal set of measure zero and most fluid elements escape from the mixing region in a finite time. Thus the asymptotic Lyapunov exponent for these transiently chaotic trajectories is zero. The local pointwise Hölder exponent of the concentration field for a decaying scalar injected into the mixing region is given by (6.32), where λ is non-zero only for the backward trajectories that are exactly on the unstable manifold of the chaotic saddle formed by the bounded orbits within the mixing zone. When $b > \lambda$ the concentration field is smooth everywhere, but when $b < \lambda$ cusps with Hölder exponents $\alpha = b/\lambda$ form along the unstable manifold (Fig. 6.7), that is superimposed onto a smooth background with $\alpha = 1$ (Neufeld et al., 2000). Numerical simulations show that the location of strong gradients of the concentration field coincides with the fractal manifold of the chaotic saddle.

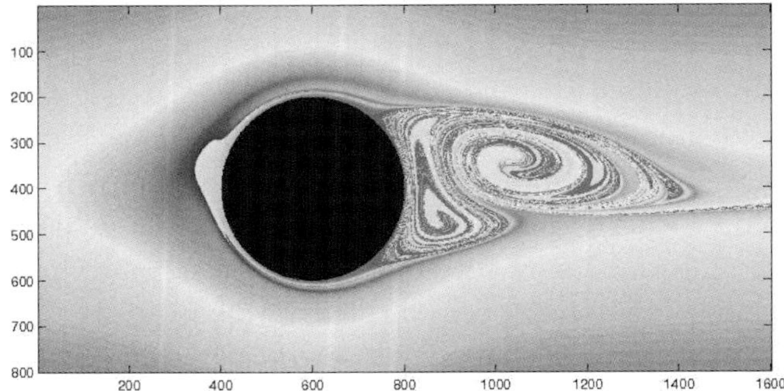

Figure 6.7: Numerical simulation of a decaying tracer in the Kármán vortex flow in the wake of a cylindrical obstacle. The flow is from left to right and a source with Gaussian distribution is located upstream on the left from the obstacle. The concentration field changes smoothly almost everywhere except for the filaments behind the cylinder that coincide with the unstable manifold of the chaotic saddle.

Although the singularities appear only on a set of measure zero, again they can contribute to the global averages and affect the scaling of the structure functions. For concentration fluctuations over a distance l the fractal set of dimension D_f in a d dimensional space contributes to the average in a proportion of

$$p_f(l) \simeq \frac{N_f(l)}{N_{total}(l)} \simeq \frac{l^{-D_f}}{l^{-d}}. \tag{6.47}$$

Thus the structure function is

$$S_q(l) \simeq p_f(l)l^{q\alpha} + [1 - p_f(l)]l^q = l^{d-D_f}l^{q\alpha} + (1 - l^{d-D_f})l^q, \tag{6.48}$$

and the scaling exponents in the $l \to 0$ limit are

$$\zeta_q = \min\left\{ q, \frac{qb}{\lambda} + d - D_f \right\}. \tag{6.49}$$

In the case of two-dimensional open flows $D_f = 2 - \kappa/\lambda$ (Eq. (2.85)), where κ is the escape rate, therefore

$$\zeta_q = \min \left\{ q, \frac{qb + \kappa}{\lambda} \right\}. \tag{6.50}$$

For small q the smooth background dominates, i.e. $\zeta_q = q$, but for higher orders, $q > q_c = \kappa/(\lambda - b)$, the scaling exponents are determined by the singular behavior along the fractal manifold if $\lambda > b$.

Resetting model

Figure 6.8: Typical structure of the concentration field in the resetting model in which the resetting is applied intermittently after each flow period. Advection is represented by the sine-flow (2.66) and the black/white resetting regions are two vertical bands located near the left/right boundaries and width equal to one tenth of the domain size. The concentration field is shown at the end of a flow period before resetting.

A different form of strong intermittency is generated when the chemical activity is localized into certain regions of the flow. In the *resetting model* introduced by Pierrehumbert (1994), fluid elements carry the chemical concentrations with no change, except when they enter certain regions where the concentration is set instantaneously to a value characteristic to that particular resetting region. Thus, the chemical concentration in a fluid element indicates its origin, or more precisely the last visited chemically active region in the past. Such process is characteristic to large scale atmospheric flows and it is also relevant for some laboratory experiments. When there are at least two such resetting regions in an ergodic chaotic flow, the chemical field (if diffusion is neglected) can only take on two possible concentration values that form a complex pattern in space as shown in Fig. 6.8. With weak diffusion the spatial structure remains roughly the same except in that the step-like discontinuities are smoothed out at very small scales.

Pierrehumbert (1994) observed numerically that the power spectrum of such concentration field is steeper than the usual k^{-1} Batchelor spectrum predicted for the passive scalar problem with additive source/sink type forcing. The resetting model can be interpreted as a strongly non-uniform chemical relaxation process (Neufeld et al., 2002c) whose relaxation rate is infinite inside the resetting regions and zero elsewhere. The pointwise Hölder exponent of the concentration field, given by Eq. (6.32), is determined by the Lagrangian chemical dynamics along the past trajectory of the fluid elements. The Lagrangian average decay rate is infinite almost everywhere, which implies $\alpha = 1$, i.e. a smooth spatial structure, except for points on the special trajectories that have never visited any of the resetting regions. For these the relaxation rate and the local Hölder exponent α are both zero. Thus analogously to the escape from the mixing regions of open flows, in a flow with ergodic chaotic advection there is a chaotic saddle with a fractal structure that contains all trajectories that never intersect any of the resetting regions as $t \to \pm\infty$. The unstable manifold of this chaotic saddle forms the boundary of fluid masses originating from different resetting regions producing sharp step-like structures in the concentration field on a fractal set. We note that qualitatively similar structures often appear e.g. in the distribution of trace gases like ozone measured along aircraft tracks, indicating boundaries between air masses that originate from different chemically active regions.

Neufeld et al. (2002c) found that the escape rate for this problem can be well approximated by $\kappa \simeq -\ln\left(\sum_i A_i\right)/A_{total}$, where A_i is the area of the chemically active resetting regions and A_{total} is the full area of the domain covered by the chaotic mixing. The fractal dimension of the set of discontinuities in the concentration field, i.e. with $\alpha = 0$, can be written as $D_f = 2 - \kappa/\lambda$ where λ is the asymptotic Lyapunov exponent. Using Eqs. (6.48)-(6.49) we obtain the scaling exponents as

$$\zeta_q = \min\{q, d - D_f\} = \min\{q, \kappa/\lambda\} \ . \tag{6.51}$$

Thus, the scaling exponents saturate at κ/λ for $q > \kappa/\lambda$. The exponent of the power spectrum is $\gamma = \zeta_2 + 1 = 1 + \kappa/\lambda$ that is steeper than the Batchelor spectrum, as observed in experiments (Williams et al., 1997).

6.5 Two-dimensional turbulence with linear damping

Nam et al. (2000) have shown that the theoretical framework introduced in the previous sections for the description of passively advected decaying scalar field also applies to the description of the small scale structure of the vorticity field in a two-dimensional turbulent flow with linear damping. The vorticity dynamics in this case is described by the Navier-Stokes equation

$$\frac{\partial \omega}{\partial t} + \mathbf{v} \cdot \nabla \omega = \nu \nabla^2 \omega - \mu\omega + f_\omega \tag{6.52}$$

where μ is a drag or friction coefficient and the forcing f_ω is a source of vorticity at large scales. This problem describes the two-dimensional turbulent flow in soap film experiments or in shallow fluid layers where the drag μ arises due to interaction with the surrounding air or friction at the bottom of the container, respectively. This is also known as Ekman friction in the context of geophysical flows.

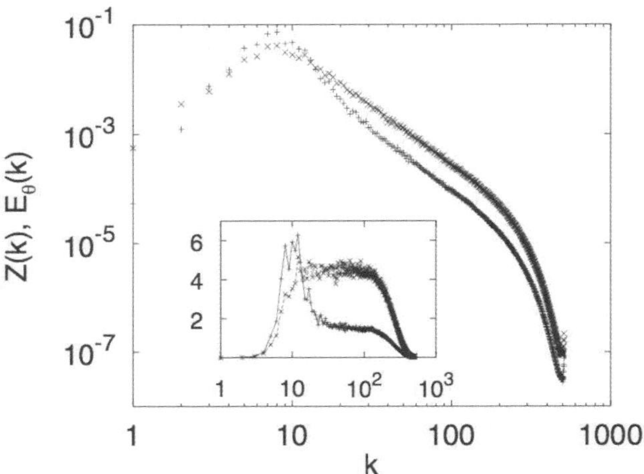

Figure 6.9: Comparison of power spectra of decaying scalar ($\times$) and vorticity ($+$) from numerical simulation of a two-dimensional turbulent flow with Ekman friction. Inset shows the ratio $Z(k)/E_\theta(k)$, which is roughly constant for large k. (From Boffetta et al. (2002))

Although Eq. (6.52) is very similar to the equation for transport of the decaying passive scalar, an important difference is that in this case the flow $\mathbf{v}$ is not independent of the transported scalar field ω. Thus, the vorticity field is a dynamically active scalar related to the flow field through the relationship $\omega = \mathbf{z} \cdot (\nabla \times \mathbf{v})$, where $\mathbf{z}$ is the unit vector perpendicular to the plane of the flow. Nevertheless, Nam et al. (2000) have shown that the friction has a similar effect on the energy spectrum as the decay in the case of the scalar spectrum. The kinetic energy spectrum is related to the enstrophy spectrum $Z(k) = \langle \omega_k^2 \rangle$ by $E(k) = Z(k)/k^2$, and therefore in the enstrophy cascade range

$$E(k) \sim k^{-3-\mu/\lambda} \tag{6.53}$$

that has a steeper spectral slope than the classical Kraichnan k^{-3} spectrum. This has been confirmed experimentally by Boffetta et al. (2005) using a quasi-two-dimensional electromagnetically driven flow in a thin fluid layer. In this experiment the friction coefficient μ, due to the bottom of the container, was measured directly through the exponential decay of the total energy when the forcing was switched off and was found to be roughly proportional to the inverse of the square of the thickness of the fluid layer. In the presence of forcing the steady state spectral slope was steeper than k^{-3} and the correction was proportional to the friction coefficient in agreement with Eq. (6.53). This may also explain the somewhat steeper than expected spectral slopes observed in some of the earlier soap film experiments (Kellay et al., 1995) where friction is due to the surrounding air. Boffetta et al. (2002) have shown numerically that the vorticity field has the same scaling properties as a decaying scalar in the same flow with scaling exponents given by (6.40) (Fig. 6.9).

Chapter 7

Mixing in Autocatalytic-type Processes

The importance of the mixing rates in nonlinear reactions, especially in autocatalytic ones (Sect. 3.1.3), or in reaction schemes with an autocatalytic step, has been recognized some time ago (Epstein, 1995). The changes of the different reaction rates and of their relative importance may induce qualitative changes in the dynamics when the stirring intensity is varied. This was nicely illustrated by Luo and Epstein (1986) who studied the bifurcation behavior of the oscillatory chlorite-iodide reaction as a function of the stirring rate. The impact of mixing in competing autocatalytic reactions was illustrated by the classical experiment of Kondepudi et al. (1990). The authors analyzed the chirality of crystals precipitating from a supersaturated solution of sodium chlorate under different mixing conditions. In each experiment performed in the absence of stirring, about half of the crystals were laevo- and half dextro-rotatory. But when mechanical stirring was fast enough, nearly all the crystals obtained in the same experiment were of the same chirality, either laevo or dextro.

The intensity of mixing also plays an important role in controlling the amplitude of stochastic fluctuations that trigger certain reactions. An example of this is described by Nagypál and Epstein

(1986) where they observed that the reaction times for the chlorite-thiosulfate reaction increase strongly with the intensity of stirring. In the parameter regime considered, the initial step of a fast autocatalytic pathway requires concentration fluctuations above a certain threshold. The stronger the stirring the more homogeneous is the concentration field, therefore the time needed to trigger the reaction increases.

In this Chapter we discuss the role of mixing in reaction systems where the chemical or biological dynamics is some kind of autocatalytic process. The main common feature of this type of reaction dynamics is that the formation of the reaction product contributes to the increase of its own rate of production. This positive feedback is usually limited by other factors. Typical examples of such system are combustion reactions, where the heat released from exothermic reactions raises the temperature and leads to enhanced burning rate. Similarly in plankton population dynamics the growth of phytoplankton results in a larger population that has an increased overall production rate of new plankton biomass. This growth will also be associated with a higher total rate of consumption of nutrients, that eventually could become a growth limiting factor.

From a dynamical point of view the systems in this class typically have multiple steady states and then the autocatalytic process is associated to transitions between them. A special case with a single steady state, that also belongs to this class, is the excitable dynamics where apart from the basic state an excited "meta-stable" state also exists, that is effectively stable for short times. In any case, late time approach to the final state after the autocatalytic dynamics is usually a decay-type process already described in the previous Chapter. The final state in most of the problems considered in this Chapter is typically a trivial spatially uniform steady state. Thus, the main focus in this Chapter will be on the transient temporal dynamics, i.e. to understand and quantify the influence of the fluid transport on the progress of the autocatalytic process. As we will show, the situation is somewhat different in open flow systems where the same type of autocatalytic processes can produce a non-trivial long-time asymptotic state with complex spatial structure.

7.1   Mixing in autocatalytic reactions

As a canonical model representing the reaction $A + B \rightarrow 2B$ (see Sect. 3.1.3) we consider the Fisher equation (4.16) with advection (Neufeld et al., 2002b)

$$\frac{\partial C}{\partial t} + \mathbf{v} \cdot \nabla C = \frac{1}{\mathrm{Pe}} \nabla^2 C + \mathrm{Da}\ C\left(1 - C\right). \tag{7.1}$$

where $C = C_B/(C_A + C_B)$ is the concentration of species A normalized by the total concentration, $C_A + C_B$, that is chosen to be initially constant in space. This property is conserved in time. $C(\mathbf{x}, t) = 0$ and $C(\mathbf{x}, t) = 1$ are homogeneous steady states of unstable and stable character, corresponding to pure A and pure B, respectively. We consider the influence of chaotic fluid flow on this dynamics in two situations: closed and open flow. To illustrate the closed flow case we use the alternating sine-flow (see Sect. 2.5), and as an example of a chaotic open flow we use the blinking vortex-sink flow from Sect. 2.6, with the inflow in the unstable $C = 0$ state, i.e. species A is continuously fed into the mixing zone representing the reactor. The results described below are obtained by integrating (7.1) using a semi-Lagrangian method and diffusion is taken to be small, so that $\mathrm{Pe} \simeq 1000$, and Da is varied from zero (no reaction or very fast flow) to several hundred.

7.1.1   The closed-flow case

We consider the time evolution of a localized perturbation of the form $C(x, y, t = 0) = C_0 \exp\left[-\left(x^2 + y^2\right)/2\sigma^2\right]$, with C_0 and σ small, on top of the unstable $C = 0$ solution. In terms of the reaction $A + B \rightarrow 2B$, this represents a diluted localized patch of the activating substance B, added to the A component. $C = 1$ corresponds to pure B and $C = 0$ to pure A. The numerical simulations show that there is a critical value of Da, in this particular case $\mathrm{Da}_c \approx 2$, at which the dynamics changes qualitatively.

    Figure 7.1 shows the case of small Da, i.e. when the reaction is slow and/or the stirring is fast. The initial perturbation diffuses quickly and while the total amount of product increases the local

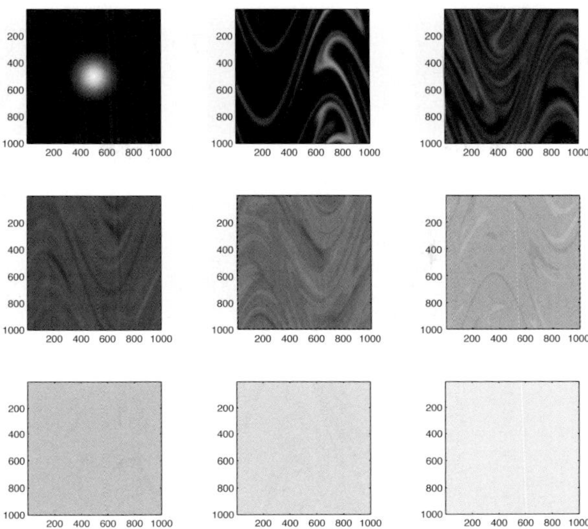

Figure 7.1: Snapshots of the spatial distribution for the autocatalytic reaction (7.1) in the alternating sine-flow (2.66) for a subcritical Damköhler number, at times $t =$ 0, 3, 6 (first row), 9, 12, 15 (second row), 18, 21, 24 (third row) with the time unit based on the period of the flow. Dark corresponds to the unreacted state $C = 0$ and light gray is the fully reacted state $C = 1$.

concentration within the perturbed region decays towards the homogeneous state $C = 0$. The whole concentration becomes nearly homogenized, and the distribution is reminiscent of the decaying strange eigenmode described in Chapter 6. This is not surprising, since in this regime C is small and the reaction dynamics can be linearized around the $C = 0$ steady state, so that the resulting problem becomes equivalent, through a simple change of variables, to the passive scalar problem. However, although spatial non-uniformities decay this homogeneous state is not stable, and after a short transient the dynamics follows a spatially homogeneous transition towards the stable steady state $C = 1$.

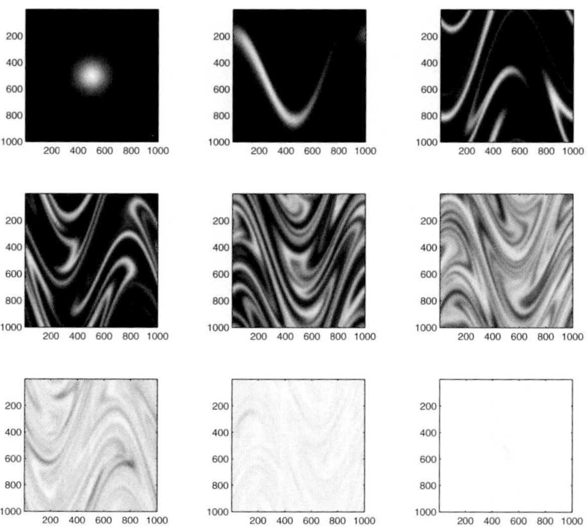

Figure 7.2: Snapshots of the spatial distribution for the autocatalytic model (7.1) in the alternating sine-flow (2.66) for a supercritical Damköhler number, at times $t =0, 1, 2$ (first row), 3, 4, 5 (second row), 6, 7 and 8 (third row).

Figure 7.2 shows the concentration evolution for $Da = 7.0 > Da_c \approx 2.0$. Now, the initial perturbation does not decay, but expands producing a filament of roughly constant width and rapidly increasing length. Since the system is closed by the periodic boundary conditions, the whole domain becomes soon filled with the growing filament and finally the system converges to the fully reacted $C = 1$ stable phase. The width of the filament increases with Da, and the average filament profile appears to be independent of the details of the initial perturbation, indicating that it is determined by the interplay between the chemical and transport dynamics. Figure 7.3 shows the time evolution of the total concentration. In all cases an initial growth as $\sim e^{Dat}$ is observed. For the smallest values of Da this growth persists and follows closely the behavior of the

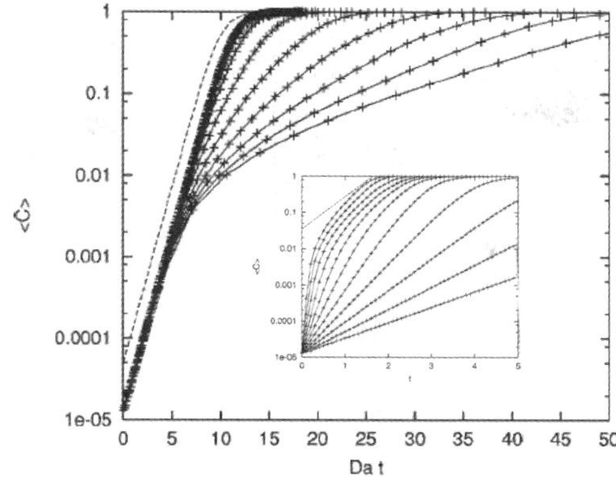

Figure 7.3: The time dependence of the total concentration as a function of the rescaled time Dat, at several values of Da for the autocatalytic dynamics in the closed sine-flow of Eq. (2.66): Da increases from top to bottom, dashed line is the homogeneous result, which is approached as Da $\to$ 0. The inset plots the evolution in terms of the unscaled time, showing that at any given time the amount of C is larger for larger Da, which corresponds to the most inhomogeneous configuration. Dashed line in the inset is the exponential growth of the length of a material line.

homogeneous system (dashed line in the main panel of the figure) until saturation at $C = 1$ is reached. For smaller Da, corresponding to faster chemistry or slower stirring, there is a crossover to a second slower exponential growth before saturation (this is best seen in the inset, where the exponential growth of the length of a passive filament is also shown).

7.1.2 The open flow case

In the blinking vortex-sink chaotic open flow system we have again two qualitatively different regimes, depending on Da, with a transition at the critical value Da$_c$ $\approx$ 2.3. As before, for small Da the initial perturbation decays fast towards $C = 0$. This is, however, not followed by an homogeneous transition to the $C = 1$ state from the $C \approx 0$ unstable configuration. The perturbation is now completely

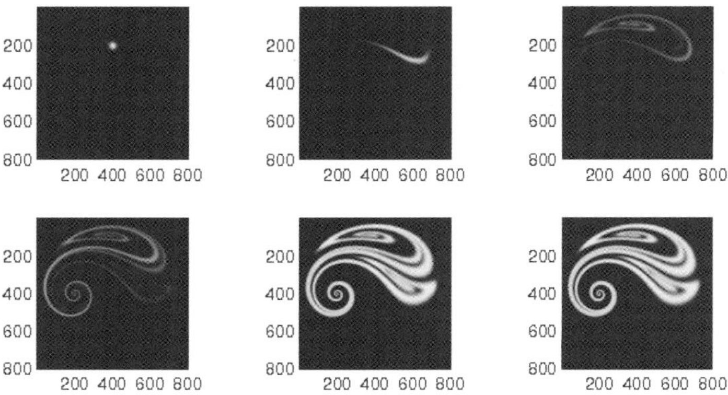

Figure 7.4: Snapshots of the spatial distribution for the autocatalytic model (7.1) in the open blinking vortex-sink flow at time intervals equal to the flow period for a supercritical Damköhler number, Da = 7.0. Note, that after a transient time a time-periodic asymptotic state is reached where the autocatalytic growth, localized on the fractal unstable manifold, is balanced by the loss of product due to the outflow from the mixing region, in this case through the point sinks.

expelled from the open system through the sinks, and the inflow boundary conditions force the system back to the $C = 0$ state, which is then established in the whole domain. When Da > Da$_c$, the initial perturbation produces a growing filament that propagates by advection. However, due to the continuous outflow of reaction products (B) and inflow of fresh reactant (A) this can not fill the whole domain. After some transient the system reaches a time-periodic asymptotic state (with the period of the flow) in which there is a balance between the production of B and its loss due to the outflow from the open system. The spatial evolution is displayed in Fig. 7.4 that shows the reaction product gradually tracing out the unstable manifold of the chaotic saddle of the flow. The product distribution produces a fattened up version of the manifold, covered by filaments of finite width, so that the small scale structure below this characteristic scale is lost. The filament width increases with Da.

Figure 7.5 shows the final total concentration (that fluctuates periodically in time following the flow oscillations, therefore the time average is plotted) clearly indicating that a continuous transition occurs at Da_c.

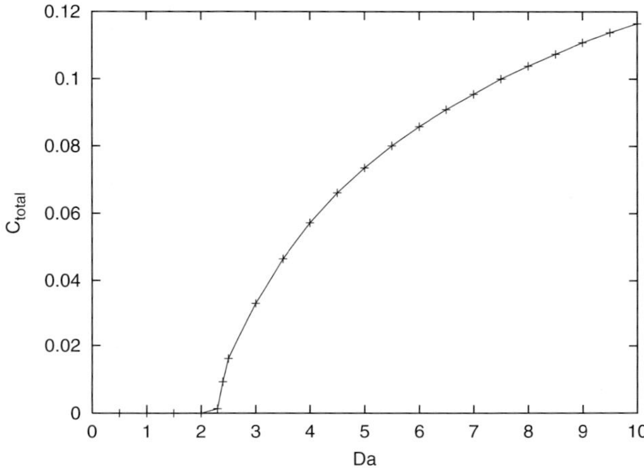

Figure 7.5: Total amount of reaction product $\langle C \rangle$ vs Da for the autocatalytic reaction advected in the alternating vortex-sink open flow system in the stationary state at a given fixed phase of the periodic flow. Note that a transition takes place at Da ≈ 2.2, below which the escape dominates and the autocatalytic component is washed out from the system. Above the critical value a non-zero steady state exists as a result of a dynamical equilibrium between the reaction and the outflow.

The observations described above are not specific to the blinking vortex-sink flow, nor specifically tied to the reaction-advection-diffusion modelling framework. Toroczkai et al. (1998) and Károlyi et al. (1999) have also observed the same phenomenon in numerical simulations using a cellular automata type model of the autocatalytic reaction and advection by an open flow created in the wake of a circular obstacle. Again, it was found that the replicating species accumulates along filaments that trace out the unstable manifold of the chaotic saddle of the corresponding advection dynamics (Fig. 7.6). In this type of models, in which the reaction takes place at discrete times, an *emptying* transition leading to the $C = 0$ state also occurs when the reactivity is decreased or for increasing escape rate, as the filament growth can not compensate for the loss through the outflow (Károlyi et al., 1999).

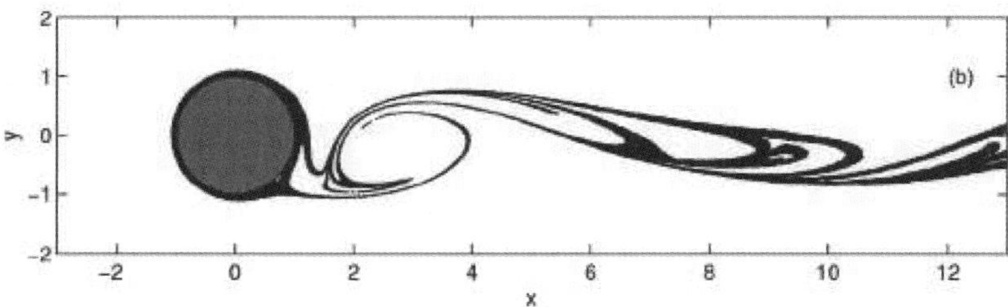

Figure 7.6: Distribution of autocatalytic particles in the time-periodic open flow in the wake of a cylinder following the unstable manifold of the advection dynamics (from Károlyi et al. (1999)).

In both the open and in the closed flow systems the main qualitative feature that determines the dynamics is the presence or absence of the filaments carrying the reaction products. Consequently, the observed behavior can be explained qualitatively by the reduced lamellar or filament model introduced in Sect. 2.7.1.

7.1.3 Results from the filament model

As discussed in Sect. 2.7.1 we can approximate the chaotic flow locally by a pure strain around an hyperbolic point that we place at the origin of the coordinate system, with the x axis aligned to the convergent direction. Note that the point in the center represents a moving hyperbolic point that is advected by the flow, and thus the corresponding strain rate can be approximated by the Lagrangian average strain, given by the Lyapunov exponent. In the y direction the concentration is quickly homogenized, so that one can focus on the filament profile $C(x, t)$ described by

$$\frac{\partial C}{\partial t} - \lambda x \frac{\partial C}{\partial x} = D \frac{\partial^2 C}{\partial x^2} + kC\left(1 - C\right). \tag{7.2}$$

Note that this local equation represents an open flow in which fluid escapes along the y direction. Nevertheless, this locally applies to any flow region where convergence of Lagrangian trajectories occurs, regardless of the globally open or closed character of the flow field. We choose boundary conditions: $C(x \to \pm\infty) = 0$, so that we focus on the concentration profile of a single isolated filament. We also require $C(x)$ to be normalizable, i.e. the quantity $C_T = \int_{-\infty}^{\infty} C(x)dx$ is finite. A localized steady filament solution of Eq. (7.2), $C_F(x)$, corresponds to a homoclinic trajectory for $x \to \pm\infty$, that forms a loop connecting the $C = 0$ fixed point to itself. Since the velocity field in (7.2) has no characteristic lengthscale nor characteristic velocity, a scaling appropriate to this case is to measure time in units of λ^{-1}, and space in units of $\sqrt{D/\lambda}$. In these units

$$\frac{\partial C}{\partial t} - x\frac{\partial C}{\partial x} = \frac{\partial^2 C}{\partial x^2} + \widetilde{\mathrm{Da}}\,C\,(1 - C) \tag{7.3}$$

where $\widetilde{\mathrm{Da}} \equiv \mu/\lambda$ is now a Lagrangian Damköhler number measuring the strength of chemical dynamics with respect to the Lagrangian average strain rate of the original flow. Note that the diffusion coefficient has disappeared, being absorbed in a simple rescaling of space, and the Péclet number does not appear in the filament equation. However, a Lagrangian Péclet number can be defined as the ratio of the diffusive and flow lengthscales in the two-dimensional problem as, $\widetilde{\mathrm{Pe}} = L\sqrt{\lambda/D}$, that gives the conversion factor of the lengthscale units between the one-dimensional filament equation and the original two-dimensional problem.

Numerical solutions of Eq. (7.3) quickly converge to a steady state representing the transverse profile of a concentration filament (Neufeld et al. (2002b); in the context of plankton filaments see also McLeod et al. (2002)). Examples are shown in Fig. 7.7 for different values of $\widetilde{\mathrm{Da}}$. This type of filament solutions exists for $\widetilde{\mathrm{Da}} > 1$. At large $\widetilde{\mathrm{Da}}$ the center of the filament profile has a flat plateau at the value $C = 1$, that ends at two sharp fronts connecting this region to the $C = 0$ state. When $\widetilde{\mathrm{Da}}$ is just slightly above the critical value the filament has the shape of a small Gaussian-like hump, whose height decreases to zero when the critical value is approached from above. This suggests a continuous bifurcation of the filament solution

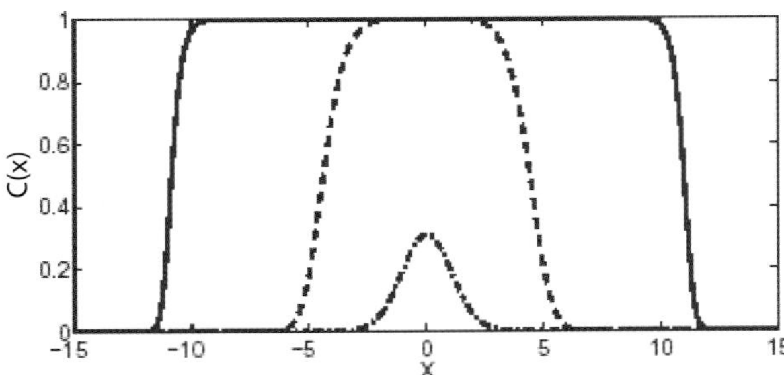

Figure 7.7: Steady state solutions of the autocatalytic filament model (7.3) for three different values of $\widetilde{\mathrm{Da}}$ (larger $\widetilde{\mathrm{Da}}$ corresponds to larger solution).

starting from the homogeneous solution $C(x) = 0$ at $\widetilde{\mathrm{Da}}_c$. To explore this possibility we first analyze the linear stability of the $C(x) = 0$ solution. By linearizing $\widetilde{\mathrm{Da}}C(1 - C) \approx \widetilde{\mathrm{Da}}C$, Eq. (7.3) becomes identical to (4.13) with $\lambda = D = 1$ and $\mu(t) = \widetilde{\mathrm{Da}}$, and the exact solution that satisfies the boundary conditions $C(x \to \pm\infty) = 0$ is of the form (4.15)

$$C(x,t) = e^{(\widetilde{\mathrm{Da}}-1)t} e^{-x^2/2} \sum_{n=0}^{\infty} C_n e^{-nt} H_n(x). \qquad (7.4)$$

When $\widetilde{\mathrm{Da}} < 1$, all the modes in the sum decay in time towards the $C = 0$ state, but for $\widetilde{\mathrm{Da}} > 1$ at least the mode with $n = 0$ grows, and $C = 0$ becomes unstable to small localized perturbations. To obtain a nonlinear approximation of the steady filament solution close to the bifurcation point $\widetilde{\mathrm{Da}}_c = 1$, a perturbation approximation can be used by introducing

$$\widetilde{\mathrm{Da}} = 1 + \epsilon \mathrm{Da}_1 + \epsilon^2 \mathrm{Da}_2 + \cdots \qquad (7.5)$$

$$C(x) = 0 + \epsilon C_1(x) + \epsilon^2 C_2(x) + \cdots . \qquad (7.6)$$

Collecting terms of order ϵ gives (the prime denotes derivative with respect to x):

$$C_1(x)'' + xC_1(x)' + C_1(x) = 0, \qquad (7.7)$$

whose bounded solution is

$$C_1 = A e^{-\frac{x^2}{2}} \tag{7.8}$$

with A an arbitrary constant, that can be determined from the equation for the terms of order ϵ^2

$$C_2'' + x C_2' + C_2 = C_1^2 - \mathrm{Da}_1 C_1. \tag{7.9}$$

Using the Fredholm alternative theorem or, equivalently, by noticing that $\int_{-\infty}^{\infty} \left(C_2'' + x C_2' + C_2 \right) dx = \int_{-\infty}^{\infty} \left(C_2' + x C_2 \right)' dx = 0$, and taking into account the boundary conditions and normalizability of $C(x)$, we find

$$0 = \int_{-\infty}^{\infty} \left(C_1^2 - \mathrm{Da}_1 C_1 \right) dx = A^2 \sqrt{\pi} - \mathrm{Da}_1 A \sqrt{2\pi} \tag{7.10}$$

from which $A = \mathrm{Da}_1 \sqrt{2}$. Finally, since (7.5) identifies the small parameter ϵ as $\epsilon = (\mathrm{Da} - 1)/\mathrm{Da}_1 + \cdots$, the filament solution close to the bifurcation point is

$$C(x) = \sqrt{2} \left(\widetilde{\mathrm{Da}} - 1 \right) e^{-\frac{x^2}{2}} + \mathcal{O} \left(\widetilde{\mathrm{Da}} - 1 \right)^2. \tag{7.11}$$

Up to this order, the filament width is controlled by the interplay between advection and diffusion: $w_F = l_D = 1$, or in the units of (7.2), $w_F = l_D = \sqrt{D/\lambda}$, as in the case of a passive scalar. But this is only valid near the bifurcation, i.e. when $\widetilde{\mathrm{Da}} - 1 \ll 1$.

Analytical approximations for the filament solution, going beyond this first-order perturbative expression can be found by variational methods (Menon and Gottwald, 2005), both in the small $\widetilde{\mathrm{Da}}$ and for the large $\widetilde{\mathrm{Da}}$ cases. In this second situation, the shape of the solutions (Fig. 7.7) can be easily understood: the two fronts are widely separated and thus one can argue that they behave independently, representing interfaces between the stable $C = 1$ and the unstable $C = 0$ states. As discussed in Sect. 4.2, when the advection term $\lambda x \partial_x C$ is absent from Eq. (7.2), the fronts have a width $l_D = \sqrt{D/\mu}$, and the state $C = 1$ would invade the state $C = 0$ at a constant speed $c = 2\sqrt{\mu D}$. Therefore, we expect that the front will stop at a position $x = w_F$ such that the propagation speed of the reaction-diffusion front is balanced by the convergent advection in the opposite direction. Thus the filament width can be estimated as $w_F \approx 2\sqrt{kD}/\lambda$, or in the units of equation (7.3), $w_F = 2\widetilde{\mathrm{Da}}^{1/2}$. This prediction becomes indeed correct for large $\widetilde{\mathrm{Da}}$. For stronger strain rates (i.e. decreasing Da) the filament becomes thinner and smaller and disappears when the distance between the two fronts is less than the diffusive width, i.e. when $w_F \simeq 2l_D$, or $\mu/\lambda = \widetilde{\mathrm{Da}} = 1$, which is exactly the transition point.

To obtain an asymptotic solution in the $\widetilde{\mathrm{Da}} \gg 1$ limit we focus on the region around the stationary front and introduce a new rescaled coordinate with the origin at the center of the front $x_0(\widetilde{\mathrm{Da}})$ (which in general will not coincide with the coordinate origin, determined by the strain flow fixed point):

$$\xi = \widetilde{\mathrm{Da}}^{\alpha}[x - x_0(\widetilde{\mathrm{Da}})], \tag{7.12}$$

and look for a solution in the form of an expansion

$$C(\xi; \widetilde{\mathrm{Da}}) = C_0(\xi) + \widetilde{\mathrm{Da}}^{-1}C_1(\xi) + \widetilde{\mathrm{Da}}^{-2}C_2(\xi) + \cdots . \tag{7.13}$$

Substituting into (7.3) and writing $x_0(\widetilde{\mathrm{Da}}) \simeq a_0\widetilde{\mathrm{Da}}^{\beta}$ with β and a_0 to be determined, gives

$$\widetilde{\mathrm{Da}}^{2\alpha}C(\xi)'' + a_0\widetilde{\mathrm{Da}}^{\alpha+\beta}\xi C(\xi)' + \widetilde{\mathrm{Da}}C(\xi)(1 - C(\xi)) = 0 \tag{7.14}$$

where balancing the leading order terms for large $\widetilde{\mathrm{Da}}$ requires the exponents to be chosen as $\alpha = \beta = 1/2$. At leading order we obtain the equation

$$C_0(\xi)'' + a_0C_0(\xi)' + C_0(\xi)(1 - C_0(\xi)) = 0. \tag{7.15}$$

This is exactly the same as the equation for a Fisher wavefront in a reference frame co-moving with the wave speed. It has solutions consistent with the boundary conditions $C(\xi \to -\infty) = 0, C(\xi \to +\infty) = 1$ for a range of values of a_0. But, as discussed in Sect. 4.2, the one with $a_0 = 2$ is dynamically preferred. Thus, the asymptotic value of the front width is $w_F \simeq x_0(\widetilde{\mathrm{Da}}) = 2\widetilde{\mathrm{Da}}^{1/2}$ as predicted above.

The threshold for the existence of the filament solution explains the two qualitatively different behaviors seen in the two-dimensional flows. When the steady filament solution exists, the reaction product accumulates along the filament, whose width is proportional to $(\widetilde{\mathrm{Da}}/\widetilde{\mathrm{Pe}})^{1/2}$, and it is stretched in the longitudinal direction until it fills the whole system (closed flow), or covers the unstable manifold of the saddle (open flow). No such process can occur when the filament solution does not exist locally, and then the autocatalytic reaction takes place as in an homogeneous system (closed flow), or dies out completely (open flow). The physical interpretation of the two regimes is that when the stirring is faster than the chemical reaction, the dispersion by the flow dominates over the local production. Thus, the spatial inhomogeneities disappear and the reaction proceeds in a spatially uniform fashion. Therefore in this reaction-limited regime the global production rate depends only on the reaction rate. In the other case, when the reaction is faster than the stirring, the dilution due to dispersion is compensated by local production, and the global production is limited by the rate of transport. The value of the critical Damköhler number for two-dimensional flows based on this interpretation of the filament model was found to be in good agreement with the numerical simulations (Neufeld et al., 2002b).

In the open flow case, the total amount of product within the mixing zone, accumulated on the unstable manifold of the chaotic saddle, can be calculated by recalling the definition of the fractal dimension D_f, that states that the number of boxes of linear size w needed to cover the fractal set scales as w^{-D_f}. Multiplying by the area of these boxes we get $C_T \approx w^{2-D_f}$ or $C_T \approx \left(\widetilde{\mathrm{Da}}/\widetilde{\mathrm{Pe}}\right)^{(2-D_f)/2}$. Similarly the length of the reaction interface between $C = 0$ and $C = 1$ where new product forms is $\mathcal{L} \approx \left(\widetilde{\mathrm{Da}}/\widetilde{\mathrm{Pe}}\right)^{(1-D_f)/2}$. These scaling laws are approximately satisfied at large $\widetilde{\mathrm{Da}}$ and $\widetilde{\mathrm{Pe}}$, and when $\widetilde{\mathrm{Pe}} \gg \widetilde{\mathrm{Da}}$ that is required for filaments with width much smaller than the characteristic length of the flow, that allows for a fractal pattern to develop over a significant range of lengthscales. Note the complete analogy with the fractal behavior of the binary-reaction case discussed in Sect. 5.3. Effective equations such as (5.23) describing the anomalous kinetics have also been developed (Tél et al., 2005).

Non-perturbative approach for the filament model

An alternative approach to analyze the solutions of the filament equation was developed by Menon and Gottwald (2005) and Cox and Gottwald (2006) based on a non-perturbative technique. In order to find an approximate solution close to the bifurcation point or for large $\widetilde{\mathrm{Da}}$ a test function with a few free parameters is chosen, that is suitable to provide a good approximation for the solution in some particular regimes. For example, close to the bifurcation point the solution is a bell-shaped function and can be approximated by a test function of the form

$$C(x) = A_0 \hat{C}(\eta), \quad \text{with} \quad \eta = qx \tag{7.16}$$

characterized by the amplitude, A_0, and inverse width, q. These parameters can then be determined by minimizing the error made by the restriction of the approximate solutions to the subspace of the test functions. The filament equation is projected to the subspace spanned by $\partial C/\partial A_0 = \hat{C}(\eta)$ and $\partial C/\partial q = \eta \hat{C}(\eta)'/q$ by setting to zero the functional scalar product of the filament equation with these basis functions. This results in algebraic equations for the width and amplitude and also yields the critical Damköhler number. For the autocatalytic reaction we obtain

$$\langle q^2 \hat{C}'' + \eta \hat{C}' + \widetilde{\mathrm{Da}} \hat{C}(1 - A_0 \hat{C}) | \hat{C} \rangle = 0, \tag{7.17}$$

$$\langle q^2 \hat{C}'' + \eta \hat{C}' + \widetilde{\mathrm{Da}} \hat{C}(1 - A_0 \hat{C}) | \eta \hat{C}' \rangle = 0 \tag{7.18}$$

where the prime denotes derivative with respect to η and the brackets indicate integration over η in the range $-\infty$ to $+\infty$. After some algebra, this leads to the expression for the amplitude

$$A_0 = \frac{3}{5\widetilde{\mathrm{Da}}\langle \hat{C}^3 \rangle} \left[\langle \hat{C}^2 \rangle \left(\frac{4\widetilde{\mathrm{Da}} - 1}{2} \right) - \langle \eta^2 \hat{C}'^2 \rangle \right]. \tag{7.19}$$

Using a Gaussian test function of the form $\hat{C} = \exp(-\eta^2)$, which is a good approximation close to the bifurcation point, simplifies this to

$$A_0 = \frac{3\sqrt{6}}{5} \left(\frac{\widetilde{\mathrm{Da}} - 1}{\widetilde{\mathrm{Da}}} \right) \tag{7.20}$$

and gives $\widetilde{\mathrm{Da}}_c = 1$, and for the width we obtain

$$w = q^{-1} = \sqrt{10/(7 - 2\widetilde{\mathrm{Da}})}.$$

This agrees with the perturbative solution at the transition point.

7.1.4 Front propagation in cellular flows

The autocatalytic reaction considered in the previous section was assumed to take place in a bounded system whose size is comparable to the characteristic lengthscale, L, of the flow field. In a large, or unbounded, system we can also consider the qualitatively different regime of propagation of a reaction front over distances much larger than L. This regime is probably relevant to describe observed plankton waves such as the ones described by Srokosz et al. (2004). Advection distorts the front and this increases the active region where the reaction takes place and speeds up the rate of transport of fresh unreacted material into the front region. Thus, the flow enhances the front propagation speed in comparison to the speed of the purely reaction-diffusion front produced in a non-moving medium, that is $v_0 = 2\sqrt{\mu D}$.

This problem was studied by Abel et al. (2001, 2002) in the case of simple two-dimensional cellular flows defined by a streamfunction of the form

$$\psi(x, y) = \frac{UL}{2\pi} \sin\left(\frac{2\pi x}{L}\right) \sin\left(\frac{2\pi y}{L}\right). \tag{7.21}$$

We can again define the Damköhler number as $\mathrm{Da} = \mu L/U$. (Note that here there is no Lagrangian Damköhler number since in the steady cellular flow the Lyapunov exponent is zero.) In the slow reaction regime, $\mathrm{Da} \ll 1$, the front is broadly distributed over distances larger than the size of a single cell and the concentration field is almost homogeneous within a single cell. Since the reaction is slow the transport can be characterized separately by an effective diffusivity, $D_{eff} = D\mathrm{Pe}^{1/2}$ valid for $\mathrm{Pe} \gg 1$ (see Sect. 2.2.2). Replacing the advection by this effective diffusion the front speed is obtained as

$$v_F = 2\sqrt{\mu D_{eff}} = v_0 \mathrm{Pe}^{1/4}, \text{ for } \mathrm{Da} \ll 1 \text{ and } \mathrm{Pe} \gg 1. \tag{7.22}$$

In the fast reaction case, the time needed for the reaction to be completed within a single cell is limited by advection, therefore the invasion of a cell can be described by an effective reaction rate $\mu_{eff} = U/L = \mu/\text{Da}$, and the front speed is

$$v_F = 2\sqrt{\mu_{eff} D_{eff}} = v_0 \text{Pe}^{1/4} \text{Da}^{-1/2}, \text{ for Da} \gg 1 \text{ and Pe} \gg 1.$$
(7.23)

Thus, in this case the front speed enhancement due to the flow is less efficient since the advection is unable to supply fresh reactants to the front at the maximum rate to keep up with the fast reaction. In terms of the flow velocity, in these two cases the front velocity scales as $v_F \sim U^{1/4}$ for slow reactions and $v_F \sim U^{3/4}$ when the reaction is fast. These theoretical predictions have been confirmed by numerical simulations (Abel et al., 2001, 2002).

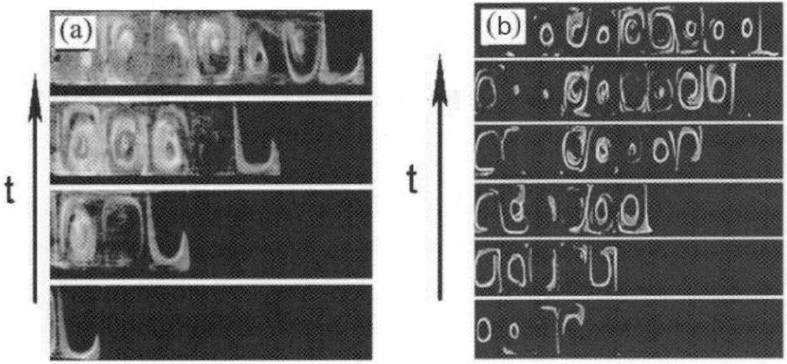

Figure 7.8: Propagation of BZ reaction fronts in an oscillating cellular flow. Images are shown every flow oscillation period. Left: Phase-locked front in which the concentration pattern at the front edge repeats every oscillation period. Right: example of an unlocked front. From Paoletti and Solomon (2005b).

In the so called geometric optics limit when the reaction front is very thin and can be approximated by a sharp discontinuity, the front speed was found to be $v_F \sim U$ with a logarithmic correction (Cencini et al., 2003b). In this limit the case of time-periodic flow was also studied. As a result of the interplay between the spatial periodicity and temporal oscillations of the flow the dependence of the front propagation speed on the flow frequency exhibits a frequency locking phenomenon. The front propagates an integer number of cells in an integer number of flow periods, so that

$$v_F = \frac{N}{M}\omega L \tag{7.24}$$

where N and M are positive integers and ω is the frequency of the time-periodic flow. This behavior with the corresponding Arnold tongue structures in the parameter space has been observed by Paoletti and Solomon (2005a) in experiments with BZ reaction fronts (Fig. 7.8).

7.2 Mixing and bistable dynamics

The behavior of a bistable chemical reaction system under mixing in many respects is quite similar to the autocatalytic case. A standard example is the model studied by Neufeld et al. (2002b), Menon and Gottwald (2005), or Cox and Gottwald (2006), given by

$$\frac{\partial C}{\partial t} + \mathbf{v} \cdot \nabla C = D\nabla^2 C + kC\left(\alpha - C\right)\left(C - 1\right) \tag{7.25}$$

or in non-dimensional form

$$\frac{\partial C}{\partial t} + \mathbf{v} \cdot \nabla C = \frac{1}{\mathrm{Pe}}\nabla^2 C + \mathrm{Da}C\left(\alpha - C\right)\left(C - 1\right) \tag{7.26}$$

which without the advection term is the same as the example discussed in Sect. 4.3. Here we have two homogeneous linearly stable states, $C = 0$ and $C = 1$, separated by an unstable homogeneous state, $C = \alpha$. In the absence of advection, a front connecting the states $C = 0$ and $C = 1$ will advance in the direction of $C = 0$ if $\alpha < 0.5$ (see Eq. (4.27)) indicating that the $C = 1$ is the dominant stable state. To be specific, we will chose $\alpha < 0.5$ and initialize the system in the $C = 0$ *less stable* state (choosing $\alpha > 0.5$ or initializing the system in $C = 1$ always leads to the disappearance of the perturbation, since in those cases the background state is the dominant one). A localized perturbation in the autocatalytic case triggers the development of the $C = 1$ state. In the absence of advection we know (Sect. 4.3) that here this will only happen for a sufficiently large perturbation, since the $C = 0$ state is linearly stable, and unstable localized solutions play the role of separating the different outcomes. We will see how this picture is modified when the advection term is included.

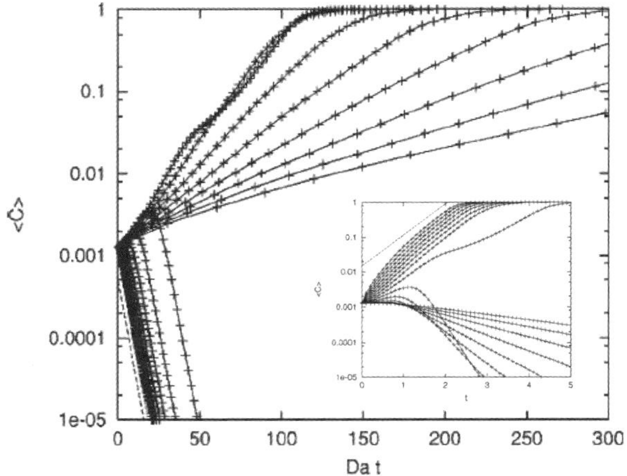

Figure 7.9: The time dependence of the total concentration as a function of the rescaled time Da t, at several values of Da for the bistable dynamics in the closed flow: Da increases from top to bottom, dashed line is the homogeneous result, which is approached as Da $\to 0$. The inset plots the evolution in terms of the unscaled time. Dashed line in the inset gives the exponential length growth of a material line in the same flow.

In the closed flow case we obtain again two different behaviors above and below a critical value Da_c. When $\mathrm{Da} > \mathrm{Da}_c$ and the localized perturbation added to the $C = 0$ state is large enough the phenomenology is similar to the autocatalytic case and configurations similar to those in Fig. 7.2 are obtained: the reaction product gradually invades the whole domain in form of a filament. The maximum concentration at the center of the filament does not vanishes smoothly as Da_c is approached from above, but the filament solution disappears suddenly from a finite non-zero amplitude. When $Da < Da_c$, or for small initial localized perturbations, rapid homogenization and a decay to $C = 0$ occurs. The concentration field looks similar to the initial stages shown in Fig. 7.1 for the autocatalytic case, but without the later homogeneous buildup of the $C = 1$ state.

Figure 7.9 shows the total concentration as a function of time. As in the autocatalytic case, the total product increases exponentially and becomes independent of Da for large Da (see inset), being close to the exponential growth rate of a material line. For small Da the decay occurs at a rate proportional to Da, as can be seen in the figure. The decay rate in this case can be easily understood because the system is rapidly homogenized and approaches the $C = 0$ state, so that the spatial derivatives can be neglected in (7.26) and the reaction dynamics can be linearized around $C = 0$, that gives

$$\dot{C}_T \approx -\alpha Da C_T \qquad (7.27)$$

from which we recover the behavior $C_T \sim e^{-\alpha Da t}$ seen in Fig. 7.9 for $Da < Da_c$.

In the open flow case, where now the $C = 0$ state is maintained by the fluid entering through the boundaries, we also find a transition at Da_c. Below the transition the perturbation is quickly diluted and the homogeneous state $C = 0$ is reached. For $Da > Da_c$ the growing filament develops until it covers the unstable manifold of the chaotic saddle, producing spatial structures similar to the ones obtained in the autocatalytic case. Figure 7.10 shows the total concentration established in the system as a function of Da. The large Da behavior is similar to the autocatalytic case, but the extinction transition at Da_c is discontinuous.

In summary, the bistable system shows a scenario quite similar to the autocatalytic case. The main difference is the discontinuous character of the transition between the two regimes, associated with the discontinuous behavior of the observed filament at Da_c, and the existence of a threshold perturbation needed to initiate it. All these features can also be recovered from the filament model (Eqs. (7.2)-

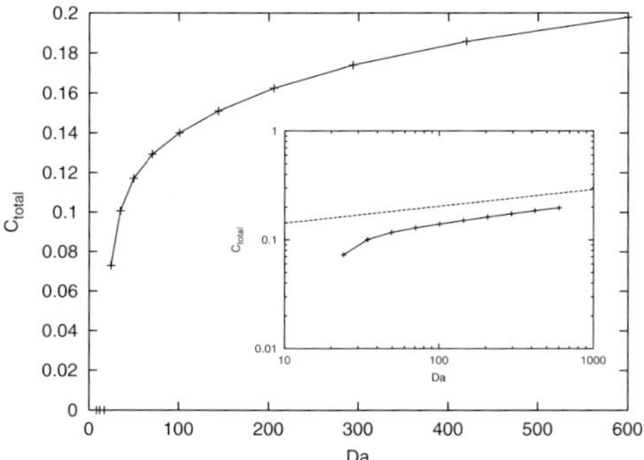

Figure 7.10: Total average concentration C_{total} in the stationary state vs Da obtained numerically for the bistable model in the open vortex-sink flow. Note the discontinuous jump to $C_{\text{total}} = 0$ at Da $=$ Da$_c \approx 24.2$, that is characteristic to the bistable dynamics.

(7.3)) with the autocatalytic term replaced by the bistable one). At variance with the autocatalytic case, here there are two branches of solutions. One of them, as expected, consists of two stationary fronts joined face to face. In the absence of flow they will move in opposite directions, but the convergent flow stops them, producing a steady state configuration. Solutions in the other branch have a lower amplitude and width, and they arise from the unstable localized solution (4.28) that also exists in the absence of the flow. Variational methods (Menon and Gottwald, 2005; Cox and Gottwald, 2006) can be used to obtain approximate analytical expressions for all these solutions. In the following we just focus on their widths.

Based on similar arguments as for the autocatalytic case the width of a filament of the $C = 1$ state flanked by sharp fronts can be obtained from the balance between the strain λ and the propagation speed of the bistable reaction-diffusion front: $w_F \approx \hat{v}/\lambda$. By using the expression (4.27), $\hat{v} = (1 - 2\alpha)\sqrt{kD/2}$, we find for the stable filament solution

$$w_F = (1 - 2\alpha)\sqrt{\frac{kD}{2}} \sim \left(\frac{\widetilde{\text{Da}}}{\widetilde{\text{Pe}}}\right)^{\frac{1}{2}}. \tag{7.28}$$

The parameter dependence is the same as in the autocatalytic case, apart from the difference in the numerical factors which depend on the stability properties of the two stable states. Note that the above solution only exists for $\alpha < 1/2$.

The width of the unstable filament solution, however, is not significantly influenced by the strain rate λ. This is because it is always relatively thin and remains localized to a small region around the origin where the strain in negligible and diffusion dominates. Thus, the shape of the unstable filament can be approximated by the exact solution for the case without the flow, which is (4.28):

$$C_u(x) = C_+ - \frac{C_+ - C_-}{1 - \frac{C_-}{C_+}\tanh^2\left(\frac{x}{w_u}\right)} \tag{7.29}$$

with

$$C_\pm = \frac{2(1+\alpha)}{3} \pm \left(\frac{4(1+\alpha)^2}{9} - 2\alpha\right)^{\frac{1}{2}}, \tag{7.30}$$

and the unstable filament width is

$$w_u = 2\sqrt{\frac{D}{\alpha k}}, \quad \text{or} \quad w_u = \frac{2}{\sqrt{\alpha \widetilde{\mathrm{Da}}}}. \tag{7.31}$$

Although the unstable filament is not reached under the evolution of the system, it plays a role in the dynamics. As $\widetilde{\mathrm{Da}}$ is decreased the flanks of the stable filament solution approach each other. Close to Da_c the filament height starts to decrease, as in the autocatalytic reaction, but now it collides with the unstable filament solution and both solutions disappear in a saddle-node bifurcation at $\mathrm{Da} = \mathrm{Da}_c$. No filament solution exists for $\mathrm{Da} < \mathrm{Da}_c$. The bifurcation point can be estimated by equating the approximating expressions for the two types of filaments: $w_F = w_u$ at Da_c, or

$$\widetilde{\mathrm{Da}}_c = \frac{2\sqrt{2}}{(1-2\alpha)\sqrt{\alpha}} \tag{7.32}$$

which compares well with the observed bifurcation point in the two-dimensional flows. The discontinuous behavior observed at Da_c in both the open and the closed flows can thus be interpreted as a consequence of the discontinuous bifurcation of the filament solutions, which is due to the presence of the unstable solution between the stable filament and the uniform $C = 0$ solution.

Specific examples of such bistable behavior arise in models of combustion. In the simplest case, a two-component system, that includes the distribution of the fuel and of the temperature field, is appropriate. The reaction takes place only when the temperature is above a certain ignition temperature, with a burning rate that is an increasing function of the local temperature following the Arrhenius law (Eq. (3.43)). The burning of fuel is an exothermic reaction that increases the local temperature and results in even higher reaction rates. This leads to the autocatalytic character of the system, while the ignition temperature acts as a threshold, responsible for the bistability.

The effect of chaotic mixing in this system was studied by Kiss et al. (2003b), Kiss et al. (2003a) and Menon and Gottwald (2007). When the reaction is initiated by raising the temperature above the ignition threshold in a small region, the resulting flame can propagate in the chaotic flow as a high temperature filament. However, as in the previous cases, this is only possible when the dispersion (or convergence) rate of the flow is not too strong, otherwise the flame filament is quenched by the flow. Properties of this system can again be well captured by the filament equation, that shows a discontinuous transition corresponding to the extinction of the flame due to the mixing in the flow. An additional complexity of this problem arises from the different diffusivities of the fuel and temperature, whose ratio is characterized by the Lewis number. The changes in the shape of the filament solutions as a function of the Lewis number was studied in Kiss et al. (2003b).

7.3   Mixing in excitable dynamics

A similar, but somewhat more complex system in the family of autocatalytic-type processes is that of an excitable reaction dynamics. This requires multiple reactions with significantly different characteristic timescales. We consider excitable dynamics occurring in the same two flows as in the previous cases (Neufeld et al., 2002c), and as a specific reaction example we focus on the FitzHugh-Nagumo

(FN) system (Meron, 1992; Murray, 1993), that is a simple standard model of excitable dynamics. Although the FN model is originally not based on chemical reactions, all the qualitative aspects of the results apply to the general class of excitable systems. As discussed in Sect. 3.1.4 the FN model describes the interaction between two species, the *activator* and the *inhibitor*, of concentrations C_1 and C_2:

$$\frac{\partial C_1}{\partial t} + \mathbf{v} \cdot \nabla C_1 = \frac{1}{\text{Pe}} \nabla^2 C_1 + \text{Da}\left[C_1\left(\alpha - C_1\right)\left(C_1 - 1\right) - C_2\right]$$

$$\frac{\partial C_2}{\partial t} + \mathbf{v} \cdot \nabla C_2 = \frac{1}{\text{Pe}} \nabla^2 C_2 + \epsilon\text{Da}(C_1 - \gamma C_2) \tag{7.33}$$

where $\epsilon \ll 1$ is a small parameter, the ratio of the relative time scales of the fast activator and of the slow inhibitor dynamics. The last term in the first equation describes bistable dynamics as in the previous section. In the homogeneous system an initial C_1 perturbation above the threshold, α, switches the system locally to the $C_1 = 1$ state. However, this is not a fully stable state since the presence of C_1 induces a slow increase of the inhibitor concentration C_2 via Eq. (7.33) which finally (after a time ϵ times larger than the time for the buildup of $C_1 = 1$) through the feedback onto Eq. (7.33), reduces the value of C_1, and finally the state $C_1 = C_2 = 0$ is recovered, which is the only stable state of the system. We study how this excitation cycle is altered by the presence of the transport terms in the fluid medium (Neufeld, 2001; Neufeld et al., 2002c). As for the other types of reaction dynamics, an initial localized pulse of C_1, with $C_2 = 0$, is added to a $C_1(x) = C_2(x) = 0$ background state and we follow the decay or propagation of this perturbation.

Again, there is a main transition at a critical Damköhler number, $\text{Da} = \text{Da}_c$. For smaller values of Da the initial perturbation is quickly diluted and the activator decays to the C_1 state, as in the bistable case. The same behavior is observed when the initial perturbation is not sufficiently large. For $\text{Da} > \text{Da}_c$ the perturbation grows as in the bistable case, forming a growing filament that eventually fills the whole system (in the closed flow case), or covers the unstable manifold of the chaotic saddle (in open flows). The filament consists now of a pulse of the C_1 concentration, with a maximum close to the excited state, and accompanied by a smaller pulse of C_2. In the closed

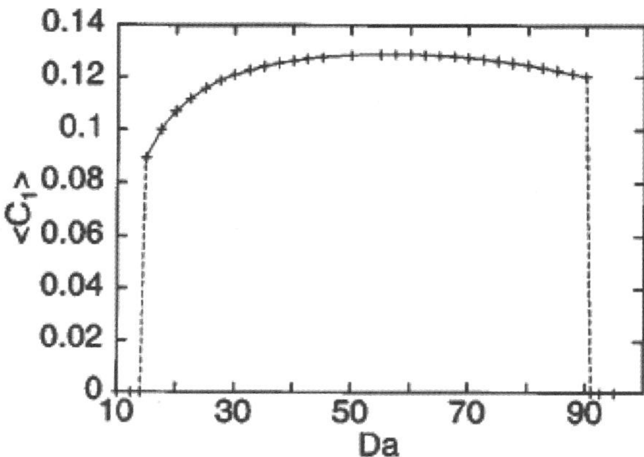

Figure 7.11: Total concentration $\langle C_1 \rangle$ in the stationary state vs Da for the FitzHugh-Nagumo dynamics in the open blinking vortex-sink flow showing two discontinuous transitions.

flow case, once the full system has reached the excited $C_1 = 1$ state, transport terms in the spatially uniform state are irrelevant and the concentrations follow the dynamics of an homogeneous system: the concentration of the inhibitor, C_2, increases until C_1 suddenly decays, and then both C_1 and C_2 return to the original steady state. This last step distinguishes the excitable dynamics from the bistable case. In the open flow, however, the final decay does not occur and as in the bistable system, the concentration distribution covering the unstable manifold of the saddle persists indefinitely. It is remarkable that the excitable cycle, which is a transient process, can be stabilized by the interaction with the open chaotic flow, in which the Lagrangian motion of the advected particles is again a transiently chaotic process.

Figure 7.11 shows the total amount of C_1 in the mixing region as a function of Da for the open flow case. The transition and subsequent behavior is similar to the bistable system. At large Da, however a second extinction transition occurs. This is not generic and depends on the geometrical structure of the flow (Hernández-García and López, 2004). What is general at large Da is a morphological change in the filament profile, in which the central part of the filament becomes a minimum of C_1, instead of a maximum, flanked by

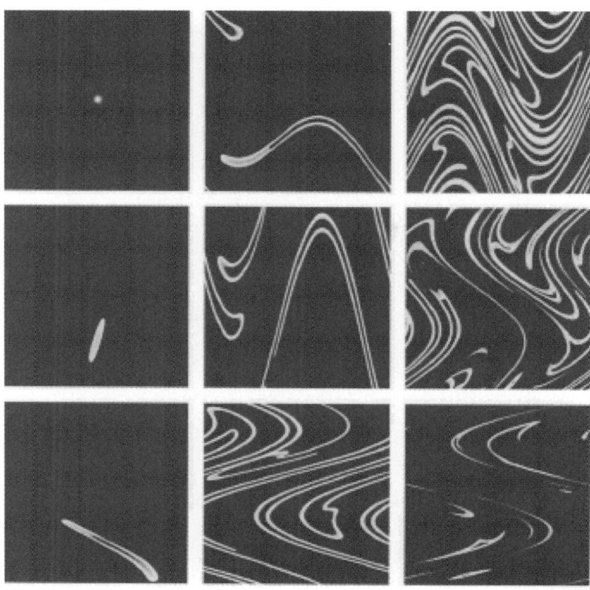

Figure 7.12: Spatial distribution of C_1 for the FitzHugh-Nagumo dynamics under the closed sine-flow (2.66). Note the double line structure of the excited filaments. (From Neufeld et al. (2002c))

two close lateral concentration maxima. The fluid in between is in the refractory state corresponding to low value of C_1 with relatively large values of C_2. This change of behavior is also observed in the closed flow case as shown in Fig. 7.12. Due to the presence of the refractory material inside the filaments the homogeneous $C_1 \approx 1$ state cannot be reached, therefore the maximum level of total excitation is smaller at these large Da values. The decay back to equilibrium is not homogeneous but proceeds by collisions and annihilations of the fronts.

These observations can be interpreted again in terms of the filament model of Sect. 2.7.1. The interesting point is that there exists a stable steady state filament solution with the excited state in the center, even though in the homogeneous system the excited state is not steady. This can be explained qualitatively by the different timescales corresponding to the dynamics of the two reaction com-

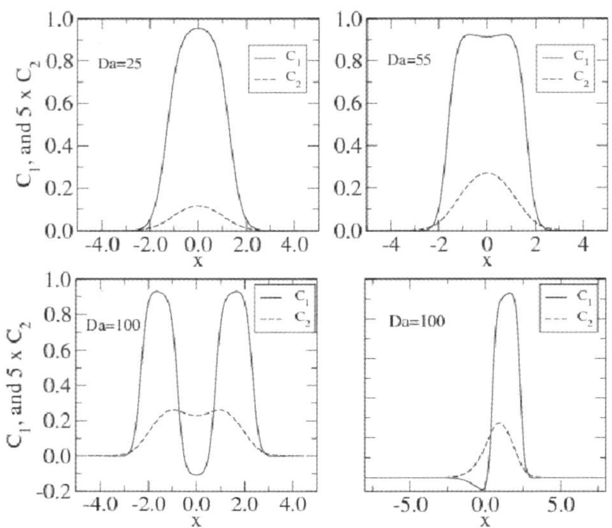

Figure 7.13: Different stable filament solutions of the excitable system (7.33), with the flow of the filament model of Sect. 2.7.1.

ponents: when the flow strain rate is slower than the fast component, C_1, but faster than the much slower inhibitor C_2, C_1 can accumulate and produce filaments, as it is produced at a faster rate than the dilution rate due to the flow. However, the slow component cannot compete with the strain. Therefore, in contrast to the standard excitable dynamics, its concentration remains low, so that the excited filament state can persist. This explains the existence of the steady excited filament solution in the one-dimensional filament model. In the closed two-dimensional system, although the inhibitor is locally diluted it will eventually build up in the system, when the excited filaments cover the entire domain. This, however, is avoided in the open flow, where some parts of the excited filaments continuously escape from the system.

Examples of stable filament solutions of the excitable model (7.33), obtained numerically for the flow of the filament model of Sect. 2.7.1, are shown in Fig. 7.13 (Neufeld et al., 2002c; Hernández-García et al., 2003). At $\mathrm{Da} = \mathrm{Da}_c \approx 12.5$ the stable filament solution collide with the unstable pulse (7.29) in a saddle-node bifurcation, so that no pulse-like solution exists for smaller Da, and a stable and an unstable branch exist for Da $>$ Da$_c$. For moderate Da, both the stable and unstable branches have C_1 profiles similar to the bistable case, only slightly modified by the presence of a concentric and much smaller pulse of C_2. Thus the bifurcation behavior, the stable and unstable filament widths, and the location of the bifurcation point is qualitatively the same as in the bistable case in this regime. This explains the first transition from extinction to filament propagation as in the bistable or autocatalytic cases. At larger Da however, instead of forming a flat plateau corresponding to the value of the excited state, the C_1 filament begins to develop a dip in its center. This arises from the interaction with the central C_2 pulse. At larger Da, new branches of filament solutions, including the asymmetric type shown in Fig. 7.13, appear in complex bifurcation scenarios (Hernández-García et al., 2003). The new extinction transition arising at large Da is associated with these new shapes, but the details depend on the way the flow produces filament collisions. For example, in an open flow in which the filaments are expelled from the chaotic region (instead of being attracted towards the sinks in the middle of the chaotic region, as in the blinking vortex-sink flow) and few collisions occur, the second transition is absent (Hernández-García and López, 2004).

7.3.1 Excitable plankton dynamics

Truscott and Brindley (1994) have shown that some plankton population models can produce excitable dynamics, and they suggested this as a possible explanation for the observed plankton blooms. The phytoplankton population plays the role of the fast component while zooplankton responds on a slower timescale to increased phytoplankton concentration. This allows for a transient plankton bloom, that can be triggered by various changes in the environment.

An interesting example of a well defined localized perturbation of the plankton ecosystem was produced by the SOIRÉE iron fertilization experiment (Abraham et al., 2000) with the aim to test the hypothesis of iron limitation of growth in regions with high nutrient, but low natural phytoplankton concentration. As it was expected, the addition of an iron compound in a patch of a few km in diameter on the Southern Ocean triggered a localized bloom within a few days, which as a consequence of the transport by the ocean flow, developed into an elongated bloom filament. Surprisingly, the bloom was still present and visible on satellite images two months after the start of the experiment as a 150 km long filament of very high plankton concentration (that could not be observed further due to cloud cover and later must have died out in the absence of light at the onset of the polar winter). The persistent bloom suggests that the expansion of the filament due to the transport by the flow was compensated by plankton growth, that was able to maintain a high concentration within the filament.

Neufeld et al. (2002a) have shown that this behavior can be explained by the interplay between excitable plankton population dynamics and chaotic flow, similarly to the excitable behavior described in the previous section. In a chaotic flow a steady bloom filament profile can be generated, that does not decay until it invades the whole computational domain as an advectively propagating bloom. The condition for the existence of the steady bloom filament solution in the corresponding one-dimensional filament model is that the rate of convergence, quantified by the Lyapunov exponent, should be slower than the phytoplankton growth rate, but faster than the zooplankton reproduction rate. In this case the phytoplankton does not became diluted by the flow and the zooplankton is thus kept at low concentration unable to graze down the bloom filament.

Hernández-García and López (2004) have considered the excitable plankton dynamics in a jet-like open flow and show the applicability of the results obtained in the excitable systems described in the previous paragraphs.

7.4 Competition dynamics

As already mentioned, mixing can also have a significant effect on competing autocatalytic reactions as observed in the distribution of chirality of crystals in the experiments of Kondepudi et al. (1990). This was reproduced in a numerical model by Metcalfe and Ottino (1994) in a system of two competing autocatalytic reactions, of the form $A + B \rightarrow 2B$ and $A + C \rightarrow 2C$ advected by a closed chaotic flow. Initially the whole system was uniformly covered with the A component and two seeds, of type B and C, were placed to different locations. The reaction was modelled in the framework of interacting particles, and was assumed to take place when the distance between two of them is smaller than a predefined reaction distance. As before, the two autocatalytic components invade the domain in form of exponentially expanding filaments. The main feature of the resulting reaction dynamics is that the outcome of the competition is extremely sensitive to the initial position of the seeds. In the final state when A is fully consumed, there is typically a dominant majority of either species B or C, but the winner cannot be predicted as even a very slight change in the initial position of one of the seeds often leeds to a completely different final composition.

In open chaotic flows, however, the behavior of competing reactions is quite different, as shown by Károlyi et al. (2000). They considered a model of competing plankton populations, represented as two self-reproducing autocatalytic species with an additional decay reaction for each component, and advection in the time-periodic flow generated in the wake of an obstacle. The reproduction and decay rates were assumed to be different for the two components. The two populations compete for a single type of nutrient, that continuously flow into the system from upstream. A model of competition is used for which the competitive exclusion principle (see Sect. 3.2.3) holds in the spatially uniform case, and one of the species completely eliminates the other when they are well mixed. Interestingly, Károlyi et al. (2000) and Scheuring et al. (2000, 2003) found that the complex filamental fractal structure created in open chaotic flows provides a mechanism through which different competitors can coexist. Although the more efficient species has a larger equilibrium population, the other competitor can also maintain a stable non-zero steady state, that would not be possible in a spatially uniform system. The mechanism is linked to the enhanced contact with nutrients of the small populations distributed on a fractal substrate, as described by the effective equations such as (5.23) or their equivalents for the autocatalytic case (Tél et al., 2005). In the context of aquatic plankton populations this phenomenon is one possible explanation for the so-called "plankton paradox" (Hutchinson, 1961), i.e. that large number of species coexist in plankton ecosystems using only a few common resources.

Chapter 8

Mixing in Oscillatory Media

There are many systems in which unsteady (e.g. oscillatory or chaotic) chemical dynamics takes place within an advected moving medium and mixing can have a significant effect on the overall global dynamics. This occurs for example in oscillatory chemical reactions in a stirred reactor. Stronger stirring leads to more uniform concentrations within the reactor, and therefore one expects that such system should be well approximated by a set of differential equations that describes the temporal dynamics of the mean concentrations, independently of the stirring rate. However, it is well known from experiments that significant non-uniformities in the concentration field persist even at high stirring rates, resulting in stirring effects. This means that certain characteristics of the mean field (e.g. period or amplitude of the oscillations) change as a function of the stirring rate (Menzinger and Dutt, 1990; Noszticzius et al., 1991). In some cases the suppression of the chemical oscillations has also been observed in a certain range of the stirring rates. Such effects cannot be captured by simple models that assume spatially uniform concentrations.

Chemical reactions in the atmosphere form a large system of interacting chemical species described by nonlinear kinetics. It has been shown that certain components of this system can also exhibit oscillatory dynamics in some range of the parameters (see e.g. Poppe and Lustfeld (1996)). When the typical period of these oscillations is within the range of the characteristic timescales of the transport processes, their interaction needs to be taken into account in models of atmospheric chemistry. The interplay between mixing and oscillations in biological systems has been studied in various contexts. Mixing plays a role in the interaction of microorganisms and cells living in a moving fluid environment, that can lead to the synchronization of oscillatory biochemical reactions (e.g. in yeast cells (Dano et al., 1999)) or can influence the outcome of cyclic competition of different bacterial strains (Kerr et al., 2002) resulting in different behavior in well mixed and static environments.

Oscillations can also arise from the nonlinear interactions present in population dynamics (e.g. predator-prey systems). Mixing in this context is relevant for oceanic plankton populations. Phytoplankton-zooplankton (PZ) and other more complicated plankton population models often exhibit oscillatory solutions (see e.g. Edwards and Yool (2000)). Huisman and Weissing (1999) have shown that oscillations and chaotic fluctuations generated by the plankton population dynamics can provide a mechanism for the coexistence of the huge number of plankton species competing for only a few key resources (the "plankton paradox"). In this chapter we review theoretical, numerical and experimental work on unsteady (mainly oscillatory) systems in the presence of mixing and stirring.

8.1 Synchronization of oscillatory dynamics by mixing

Consider the continuum description of an oscillatory medium subject to advection and diffusion

$$\frac{\partial C_i}{\partial t} + \mathbf{v}(\mathbf{r}, t) \cdot \nabla C_i = k_0 F_i(C_1, ..., C_N) + D_i \nabla^2 C_i \qquad (8.1)$$

where k_0 is a characteristic reaction rate, and we assume that the chemical dynamics described by the set of functions F_i leads to oscillatory dynamics, such that the system

$$\dot{C}_i = k_0 F_i(C_1, ..., C_N), \qquad i = 1, ..., N \qquad (8.2)$$

has a stable limit cycle asymptotic solution with period T_c/k_0, i.e.

$$C_i(t + T_c/k_0) = C_i(t) \quad \text{for large } t. \tag{8.3}$$

Here we also assume that the reaction term does not depend explicitly on the spatial coordinate, therefore the dynamics of the medium is uniform in space. It is easy to see that the spatially uniform time-periodic oscillation is a trivial solution of the full reaction-diffusion-advection system, so the question is whether this uniform solution is stable to small non-uniform perturbations and more generally, if there are any persistent spatially non-uniform solutions in which the spatial structure does not decay in time.

The qualitative behavior of the system depends on various characteristics of the velocity field. The simplest case is when there is uniform chaotic mixing in the flow over the whole domain, so that there are no transport barriers and the characteristic length-scale of the velocity field is comparable to the size of the domain. This problem has been studied by Kiss et al. (2004) using a model of the chlorine-iodine-malonic acid (CDIMA) reaction (Sect. 3.1.4) described by

$$F_1(C_1, C_2) = 1 - C_1 - \frac{4C_1 C_2}{\beta + C_1^2}; \quad F_2(C_1, C_2) = \alpha \left(C_1 - \frac{C_1 C_2}{\beta + C_1^2} \right) \tag{8.4}$$

combined with mixing by the sine-flow (Sect. 2.5) with random phase to avoid elliptic islands. Numerical simulations performed for different values of the Damköhler number show that the perturbation first produces a spatial pattern formed of filaments superimposed on the oscillating background that are stretched and folded by the chaotic advection. The contrast of the spatial pattern gradually fades and the oscillating concentration field becomes more and more uniform.

The process of synchronization of the local oscillations can be described by following the evolution of the standard deviation in space of the concentration field as a function of time (Fig. 8.1)

$$\sigma_C(t) = \sqrt{\langle C^2(\mathbf{r}, t) \rangle - \langle C(\mathbf{r}, t) \rangle^2} \tag{8.5}$$

where $\langle ... \rangle$ represents averaging over the domain. The variance σ_C^2 decays exponentially with a superimposed periodic oscillation. The

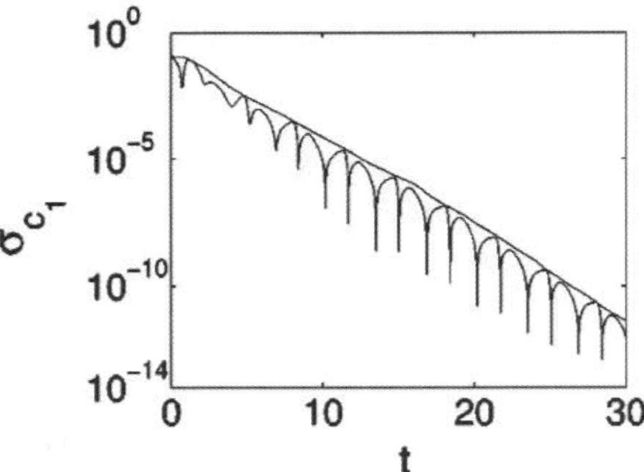

Figure 8.1: Exponential decay of the variance of the concentration fluctuations in an oscillatory reaction mixed by a chaotic flow. The smooth curve shows the decay of the passive scalar variance in the same flow, for comparison (from Kiss et al. (2004)).

decay rate was found to be independent of the reaction rate, k_0, and is the same as the asymptotic decay rate of the variance of passive scalar inhomogeneities in the same flow.

A simple explanation for this is that although the stable limit cycle is attracting, it is neutral with respect to perturbations in phase along the limit cycle as indicated by the zero Lyapunov exponent of the oscillatory system. Therefore the phase of the oscillation behaves like a passive scalar and its non-uniformity is only dissipated by mixing, that determines the asymptotic decay rate. Thus, any slow mixing can fully synchronize an oscillatory medium with a spatially uniform frequency over distances corresponding to the characteristic lengthscale of the flow (at least in the linear regime in which the full system is everywhere close to the limit cycle). Without stirring, the uniform state of an oscillatory reaction-diffusion system is unstable to small perturbations and develops into a complex pattern of propagating waves, hence stirring by a slow chaotic flow changes the asymptotic behavior of the system.

This can be shown more quantitatively for the case of a spatially localized inhomogeneity, where the stability of the uniform steady state can be analyzed using the filament equation introduced in Sect. 2.7.1. Taking the spatially uniform oscillation as a basic state and approximating the chemical dynamics by a Taylor expansion about the stable limit cycle gives

$$\frac{\partial C_i}{\partial t} - \lambda x \frac{\partial C_i}{\partial x} = k_0 \sum_j A_{ij}(t) C_j + D \frac{\partial^2 C_i}{\partial x^2} \qquad (8.6)$$

where $A_{ij}(t) = \partial F_i/\partial C_j$ are time-periodic functions with period T_c/k_0 and λ is the Lyapunov exponent of the flow. All chemicals are assumed to diffuse with the same diffusion coefficient D, which can then be set to $D = 1$ by rescaling the x coordinate by $\sqrt{D/\lambda}$. The eigenvalue problem associated with the resulting one-dimensional advection-diffusion operator of the filament model

$$\frac{\partial^2 W}{\partial x^2} + x \frac{\partial W}{\partial x} = -\alpha W \qquad (8.7)$$

has eigenmodes of the form

$$W_k = e^{-x^2/2} {}_1F_1\left(-k; \frac{1}{2}; \frac{x^2}{2}\right) \quad k = 0, 1, 2... \qquad (8.8)$$

with corresponding eigenvalues $\alpha_k = 2k + 1$. The functions ${}_1F_1$ are confluent hypergeometric functions and with k positive integer are polynomials in x^2 of order x^{2k} (this is equivalent to the Hermite polynomials in Eq. (2.91)). These eigenmodes can be used as basis functions for constructing solutions of Eq. (8.6) in the form

$$C_i = \sum_{k=0}^{\infty} U_{i,k}(t) W_k(x). \qquad (8.9)$$

Substituting into Eq. (8.6) gives

$$\frac{dU_{i,k}}{dt} = k_0 \sum_j U_{j,k} A_{ij}(t) - (2k + 1)\lambda U_{i,k}. \qquad (8.10)$$

With the substitution $U_{i,k} = \exp[-(2k + 1)\lambda t]\hat{U}_{i,k}$ we obtain

$$\frac{d\hat{U}_{i,k}}{dt} = k_0 \sum_j A_{ij}(t)\hat{U}_{j,k} \qquad (8.11)$$

which is precisely the equation that determines the set of Floquet exponents, ρ, of the limit cycle of the homogeneous case. It has solutions of the form $\hat{U}(t + T_c/k_0) = e^{\rho T_c/k_0}\hat{U}(t)$, hence the solutions of Eq. (8.10) satisfy

$$U_{i,k}(t + T_c/k_0) = e^{[\rho - (2k+1)\lambda]T_c/k_0}U_{i,k}(t) \qquad (8.12)$$

Since the limit cycle is invariant with respect to a phase shift, one of the Floquet exponents is zero and the asymptotic stability of the limit cycle implies that the other Floquet exponents have all negative real parts. Thus the amplitude of all modes decay in time and the uniform oscillatory state is stable to small localized perturbations (within the filament model and in the linear regime to which Eq. (8.6) applies). The slowest decaying mode (i.e. for $\rho = 0$ and $k = 0$) decays on average as $\simeq \exp(-\lambda t)$, that is the same as the decay rate of a passive scalar inhomogeneity in the filament model.

Going beyond the periodic oscillations case, a more complex situation is that of synchronization of a chaotic local dynamics in a mixing flow, as was considered by Straube et al. (2004). When the flow is time-independent one can look for solutions of the form

$$\mathbf{C}(\mathbf{x}, t) = \mathbf{C}_0(t) + \sum_{\mu} \mathbf{U}_{\mu}(t)\Phi_{\mu}(\mathbf{x}) \qquad (8.13)$$

in which the temporal and spatial dependence of the perturbation field are factorized. The spatial structure can be constructed from the eigenmodes of the advection-diffusion problem

$$D\nabla^2\Phi_{\mu} - (\mathbf{v} \cdot \nabla)\Phi_{\mu} = -\alpha_{\mu}\Phi_{\mu} \qquad (8.14)$$

where the eigenvalues α_{μ} are the decay rates of spatial non-uniformities due to mixing. As discussed in Sect. 2.7.2 the dominant mode is the most slowly decaying eigenmode corresponding to the smallest non-zero eigenvalue α_1 (since $\alpha_0 = 0$ corresponds to a spatially uniform eigenmode and does not contribute to the decay of non-homogeneous perturbations).

The evolution of a small non-uniform perturbation can be described by (8.13), which leads to the equation for the amplitude of the dominant eigenmode

$$\frac{d\mathbf{U_1}}{dt} = -\alpha_1\mathbf{U_1} + k_0\mathbf{A}(t)\mathbf{U_1}. \qquad (8.15)$$

With the substitution

$$\mathbf{U_1}(t) = e^{-\alpha_1 t}\hat{\mathbf{U}}_1(t) \qquad (8.16)$$

we obtain

$$\frac{d\hat{\mathbf{U}}_1}{dt} = k_0 \mathbf{A}(t)\hat{\mathbf{U}}_1 \qquad (8.17)$$

which is the equation for the linear perturbations of the homogeneous reaction problem. In the periodic oscillations case it has solutions of the Floquet type, but more generally the solution behaves at long times as a sum of exponential terms with the exponents given by the Lyapunov exponents of the chemical dynamics. If the reaction dynamics is chaotic the largest Lyapunov exponent $k_0\lambda_1^C$ is positive. Combining with Eq. (8.16), the transverse Lyapunov exponent characterizing the evolution of small non-homogeneous perturbations with zero mean is

$$\lambda_\perp = k_0\lambda_1^C - \alpha_1 \qquad (8.18)$$

that can be either positive or negative, and changes sign at the critical value of the reaction rate

$$k_0^* = \frac{\alpha_1}{\lambda_1^C} \qquad (8.19)$$

that determines the transition between non-uniform spatiotemporal patterns and purely temporal spatially homogeneous dynamics. $\alpha_1 = \alpha_1(\text{Pe})$, as determined from the solution of the eigenvalue problem (8.14).

The analysis above refers to time-independent velocity fields and equal diffusion coefficients for all species, but Straube et al. (2004) have shown that similar behavior applies to mixing in time-dependent chaotic flows. It was shown numerically that $\lambda_\perp$ is a linear function of the reaction rate and the transition in the stability of the spatially uniform state to non-homogeneous perturbations takes place when the positive Lyapunov exponent of the local dynamics is equal to the exponent describing the decay rate of the dominant eigenmode.

When the diffusion coefficients are not the same, there is a set of decay rates which are different for each component. Very different decay rates could arise, for example, when some components are not advected by the flow and the fixed and the transported components interact (e.g. some chemical species can be absorbed to the surface of a reactor, or in a marine environment sedentary species may interact with planktonic ones). In this case the transverse Lyapunov exponent cannot be expressed simply as the sum of the chemical and homogenization exponents. But the problem remains qualitatively the same and the transverse Lyapunov exponent is in general a nonlinear function of k_0 and of the set of decay rates of all the components.

8.1.1 Persistent patterns in uniform medium

The analysis above has shown that, within the linearized regime and in a fully chaotic flow, the decay of the spatial inhomogeneities due to advection and diffusion decouples from the temporal dynamics of the spatially uniform system, resulting in a simple superposition of the two behaviors. However, further experimental and numerical studies have shown that complete homogenization may not always take place in stirred oscillatory systems, even if the reactive medium is completely uniform in space. Non-uniform concentrations may persist when somewhat different flow configurations are considered in which there is no uniform chaotic mixing over the whole reactive medium. Nugent et al. (2004) studied the spatial patterns produced by the Belousov-Zhabotinsky reaction in experiments where the reactive medium was stirred by a blinking vortex flow generated by magneto-hydrodynamic forcing in a thin quasi-two-dimensional fluid layer. This flow (the closed version of the one presented at the end of Sect. 2.6) can be approximately described by two point vortices that are switched on and off alternately with period T, each having a velocity field

$$v_{x1,2} = -\Gamma \frac{y}{(x \pm b)^2 + y^2}, \quad v_{y1,2} = \Gamma \frac{x}{(x \pm b)^2 + y^2} \tag{8.20}$$

where $2b$ is the distance between the vortex centers and Γ is the vortex strength. The extent of the chaotic mixing region depends on the non-dimensional parameter $\mu = \Gamma T/b^2$. For small μ the flow is mostly regular, similar to a steady flow with both vortices switched-on simultaneously. As μ is increased more transport barriers break up and the chaotic mixing regions became larger. When μ is larger than a critical value, a bounded contiguous well mixed region develops that is filled with chaotic trajectories.

To study the effects of changing the stirring rates on the spatial patterns, the period of the vortex system was varied in such a way that the vortex strength Γ was also adjusted to keep μ constant. This procedure maintains the character and topology of the flow, since the trajectories of the fluid elements remain unchanged. In the BZ reaction the evolution of the concentration field can be easily detected by changes in color during the oscillations. In the absence of mixing the oscillatory reaction produces a complex structure of waves and spirals. For fast mixing, synchronized oscillations develop within the chaotic region of the flow. But, when the mixing is slow the spatial structure was found to remain non-uniform even inside the chaotic mixing region.

The transition between the two regimes takes place when the characteristic timescale of chaotic mixing (e.g. the inverse Lyapunov exponent) is comparable to the characteristic decorrelation time of the of the local oscillations, that was found to be significantly larger than the oscillation period of the reaction.

The apparent contradiction between these experiments and the theoretical and numerical results in the previous Section can be resolved by taking into account that in this case the chaotic mixing region does not cover the whole domain of the reaction-diffusion system. Outside the chaotic region, the flow is mostly composed of concentric transport barriers with only slow diffusive transport across them. Therefore in this region non-uniformities of the chemical concentration can persist and develop oscillatory wave patterns aligned along the transport barriers. This external region where spatial patterns do not decay acts as a persistent source of inhomogeneity, that can prevent synchronization even inside the chaotic region when the chaotic mixing is not sufficiently strong.

Non-decaying patterns in a similar stirred oscillatory system were also found numerically by Pérez-Muñuzuri (2006) in the weakly chaotic flow regime of the blinking vortex flow, i.e. for small μ. When the distance between the vortices is large, the flow has little effect on the spatial structure and a pattern of spiral waves forms as in the

static oscillatory medium. As the distance between the two vortices
is reduced, the weakly chaotic flow produces circular target patterns
whose centers coincide with the vortex centers. Note that in this case
the chaotic mixing is restricted to thin layers separated by transport
barriers due to KAM tori. Thus, the reaction-diffusion waves produced by the chemical reaction simply follow the spatial structure
of the transport barriers (Fig. 8.2). Note that such KAM tori were
not present in the random sine-flow used by Kiss et al. (2004) and
Straube et al. (2004).

Figure 8.2: Patterns generated in numerical simulations of an oscillatory reaction-advection-diffusion system (from Pérez-Muñuzuri
(2006)). The flow is composed of two point vortices that are switched-
on alternatingly. The figures are snapshots of the concentration field
for different values of the distance between the two vortex centers (decreasing from left to right).

Another set of experiments, by Paoletti et al. (2006), investigated
the synchronization of stirred oscillatory BZ reaction over distances
larger than the characteristic lengthscale of the flow. In this experiment the flow was composed of an annular ring of counter-rotating
vortices with a superimposed additional oscillatory azimuthal flow.
A simplified model of the corresponding velocity field can be written
as

$$v_x = -U\cos(kx_s)\sin(ky), \quad v_y = U\sin(kx_s)\cos(ky) , \qquad (8.21)$$

where the vortices form a linear chain along the x axis, with periodic
boundary conditions. The size of the vortices is π/k and the vortices
move along the cyclic x direction according to

$$x_s = x + \frac{v_0}{\omega}\sin(\omega t) + v_d t \qquad (8.22)$$

where ω and v_0 are the frequency and amplitude of the oscillation,
while v_d is a drift velocity and the dependence on x fixes the initial

phase of the pattern. In the absence of the oscillatory component $(v_0 = 0)$ the flow is time-independent in a co-rotating reference frame, and therefore advection is non-chaotic occurring either along closed orbits within the vortices or on open orbits along a jet between the vortices. When $v_0 > 0$ there are regular and chaotic regions separated by transport barriers.

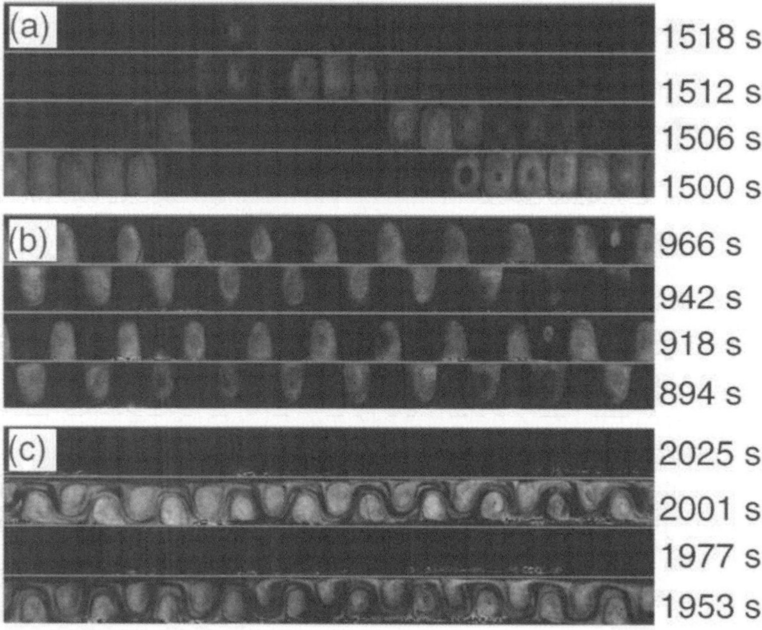

Figure 8.3: Experimental synchronization patterns in the oscillatory Belousov-Zhabotinsky reaction in a cellular flow. The horizontal direction is along an annulus, so that there are periodic boundary conditions at the ends of the images. (a) Phase waves. (b) Co-rotating synchronization. (c) Global synchronization. From Paoletti et al. (2006).

Typically, large scale transport in flows with a finite correlation length of the velocity field is diffusive with an effective diffusion coefficient D_{eff} (Sect. 2.2.2). Therefore the coarse grained structure of the oscillatory reaction in this flow should be similar to a one-dimensional oscillatory reaction-diffusion system, i.e. propagating waves and no synchronization of the local oscillations on large scales.

This wavelike behavior is indeed observed in the experiments when the drift velocity is smaller than the velocity of the oscillatory part of the flow, $v_d < v_0$. However, when $v_d > v_0$ the oscillations synchronize over the whole system consisting of about 20 vortices. Two types of coherent oscillatory modes are observed depending on the flow parameters: "corotating" synchronization when even and odd cells synchronize independently with arbitrary phases, and global synchronization when the BZ reaction oscillates in synchrony in every cell (Fig. 8.3).

The authors explain this surprising coherent behavior over length scales much larger than the characteristic scale of the flow by the superdiffusive character of the long-range advective transport. The spreading of a small cloud of tracer particles advected by the flow over long time can be characterized by the growth of the variance of the tracer coordinates which is of the form

$$\langle x^2 \rangle \sim t^\alpha. \tag{8.23}$$

In the case of diffusive transport $\alpha = 1$, but certain type of flows exhibit superdiffusive behavior with $\alpha > 1$ (Sect. 2.1.2). Superdiffusive transport indicates that fluid elements undergo "Lévy flights", i.e. occasional jumps between distant regions in the flow, providing a mechanism for efficient mixing over large distances that can synchronize the oscillatory reactions. Experiments performed over a range of flow parameters have confirmed that the long-range synchronization regime coincides with the region of superdiffusive transport in the flow. The "corrotating" synchronization is found to be associated with the presence of a transport barrier in the flow that prevents mixing between adjacent vortices.

8.2   Synchronization in non-uniform medium

In the previous oscillatory systems, the local dynamics was assumed to be the same everywhere in space and the synchronization of identical oscillatory regions was studied. In many cases the oscillatory medium is not uniform. In real chemical or biological systems this can be a consequence of non-uniform external conditions, like variation of temperature, or of illumination in a photosensitive reaction, or changes in the local carrying capacity of a plankton ecosystem. This results in an explicit dependence of the local dynamics on the spatial coordinates in the reaction-advection-diffusion equation. The non-uniformity of the medium acts like an external forcing that continuously generates inhomogeneities in the concentrations similarly to a source term of a passive scalar field.

The simplest form of a spatially non-uniform oscillatory reaction dynamics is obtained by assuming that the local dynamics is composed of identical limit cycles whose frequency is non-uniform in space. In this case the reaction term can be written in the form

$$F_i(C_1, ..., C_N; \mathbf{r}) = [1 + \delta f(\mathbf{r})]\hat{F}_i(C_1, ..., C_N) , \qquad (8.24)$$

where δ is the amplitude of the inhomogeneity and $f(\mathbf{r})$ represents the spatial structure of the frequency distribution. This type of oscillatory system in the uniformly mixing chaotic sine-flow was studied by Neufeld et al. (2003). The main control parameter is the stirring rate ν defined as the inverse of the period of the flow. The spatial coherence of the oscillations can be characterized by the time dependence of the mean concentration.

In the case of strong stirring, the mean concentration has an oscillatory time-dependence indicating synchronization, but unlike the homogeneous medium, the spatial fluctuations superimposed on the oscillatory background do not disappear completely with time, but persist indefinitely. This is a consequence of the absence of a spatially uniform oscillatory solution in this problem. The non-uniform frequency continuously generates spatial variations in the chemical concentrations, which are transferred to smaller scales by advection and dissipated by diffusive mixing at small scales. The amplitude of the oscillations of the mean field increases with the stirring rate. For very fast stirring the spatial fluctuations are small and the mean field closely follows the limit cycle of a single oscillator with frequency equal to the average frequency of the medium.

When the stirring rate is below a certain threshold the synchronized oscillation disappears and the mean concentration is almost constant apart from small irregular fluctuations that are independent of the stirring rate. In this regime the mixing is too slow to strongly couple the oscillatory dynamics of the fluid parcels. Snapshots of the concentration field taken at different times are statistically equivalent, indicating the absence of spatially coherent oscillatory dynamics. At very slow stirring the concentration field is dominated by propagating phase waves similar to the unstirred oscillatory medium but the wavefronts are distorted by advection.

The degree of synchronization can be characterized by the standard deviation in time of the spatially averaged concentration

$$R(\nu,\delta) = \lim_{t\to\infty} \sqrt{\frac{1}{t}\int_0^t \langle C\rangle^2 dt' - \left(\frac{1}{t}\int_0^t \langle C\rangle dt'\right)^2}. \qquad (8.25)$$

This order parameter measures the spatial coherence of the dynamics and can be normalized by dividing it with its value corresponding to the spatially uniform oscillatory system $R_0 = R(\delta = 0), 0 < R/R_0 < 1$. The dependence of R on the stirring rate ν is shown in Fig. 8.4 for different values of δ. In the slow stirring regime R is small and is approximately constant. The slightly larger than zero value of the order parameter in the non-synchronized state is due to the finiteness of Pe. In the limit of infinite Péclet number, corresponding to an infinite ensemble of non-identical oscillators, the mean concentration should became time-independent, i.e. $R = 0$. When the stirring rate is larger than a critical value the order parameter R increases sharply and approaches R_0 in the limit of large stirring rates.

Another important control parameter that affects the behavior if this system is the degree of inhomogeneity of the medium. It is found that the critical stirring rate at which the synchronization transition takes place depends on the strength of the inhomogeneity, δ. This is because in a medium with a larger spread of the local frequencies, faster stirring is necessary to obtain the same degree of synchronization. Furthermore, when δ is sufficiently large at intermediate stirring rates a new regime appears between the synchronized and unsynchronized phases in which the small fluctuations of the mean field disappear completely. The spatial structure clearly shows that the concentration field evolves towards a spatially uniform time-independent state that coincides with the unstable steady state of the oscillatory reaction kinetics, C_i^*, i.e. the solution of the system

$$\hat{F}_i(C_1^*, ..., C_N^*) = 0. \qquad (8.26)$$

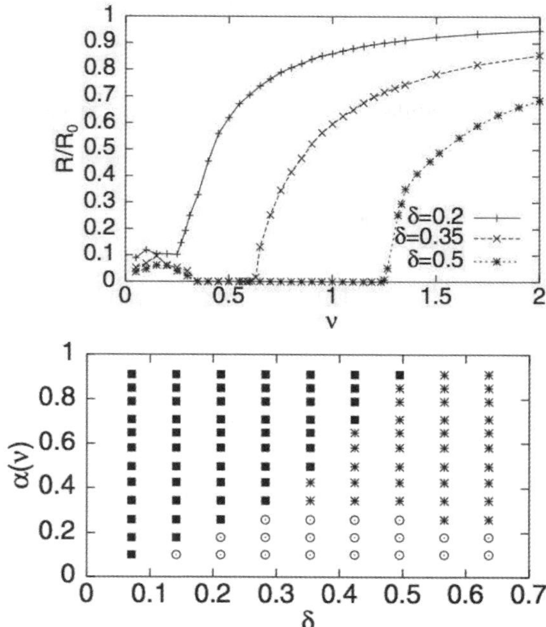

Figure 8.4: The order parameter R vs. the stirring rate shows the synchronization transition for different values of the non-uniformity parameter δ (top). Phase diagram as a function of the stirring rate ν and inhomogeneity δ (bottom). The symbols represent $\bullet$ synchronization, $*$ oscillator death and $\odot$ no synchronization (from Neufeld et al. (2003)).

Thus in the "oscillator death" state (Mirollo and Strogatz, 1990), the stirring completely inhibits the chemical oscillations, transforming the unstable steady state of the local dynamics into a stable attractor of the reaction-advection-diffusion system. This arises from the competition between the non-uniform frequencies that lead to the dispersion of the phase of the oscillations, while mixing tends to homogenize the system and bring it back to the unstable steady state. Therefore this behavior is only possible in oscillatory media with spatially non-uniform frequencies. The phase diagram of the non-uniform oscillatory system in the plane of the two main control parameters (ν, δ) is shown in Fig. 8.4.

Behavior similar to the oscillator death described above has been observed as the suppression of oscillations due to stirring in some experiments with oscillatory chemical reactions, and was also found in a few other systems without mixing, like globally coupled phase-amplitude oscillators or oscillators with delayed coupling (Mirollo and Strogatz, 1990; Ramana Reddy et al., 1998).

8.3 Noise induced oscillations in excitable media

In the previous examples, the oscillatory dynamics was an inherent property of the medium and the role of mixing was to control the degree of coherence of the oscillations. When coherent oscillations are produced, the period is only controlled by the oscillatory reaction dynamics of the medium and is not influenced by the flow. It has been observed experimentally, however, that in certain oscillatory chemical reactions the stirring rate can also change the period of the oscillations. Here we present a model that gives some insight into possible mechanisms of this type of stirring effect, based on an excitable reaction scheme subject to stochastic perturbations. The excitable system has a stable steady state, and perturbations above a certain threshold lead to a temporary large deviation from this steady state (Sects. 3.1.4 and 7.3). Such excitable dynamics is well known in a range of chemical reactions and is also relevant for the dynamics of plankton blooms. In both types of systems random perturbations naturally arise due to various external fluctuations.

A simple model for the local temporal dynamics of such systems is the FitzHugh-Nagumo model (3.59)-(3.60), written here as:

$$\dot{C_1} = C_1(a - C_1)(C_1 - 1) - C_2, \quad \dot{C_2} = \epsilon(C_1 - 3C_2) + \xi(t) \quad (8.27)$$

where $\xi(t)$ is a stochastic white noise perturbation. In general, this type of excitable systems produce an irregular sequence of pulses. The time between two consecutive pulses can be decomposed into the sum of an activation time needed by the noise to excite the system in the steady state, and an excursion time needed for return to the steady state after activation. The activation time decreases with the noise amplitude, hence for weak noise the random activation time dominates, while for large noise amplitudes the excursion time is also influenced by the perturbation. Pikovsky and Kurths (1997) have shown that in certain intermediate range of noise amplitudes, regular, almost periodic oscillations develop, a phenomenon known as "coherence resonance" (Fig. 8.5).

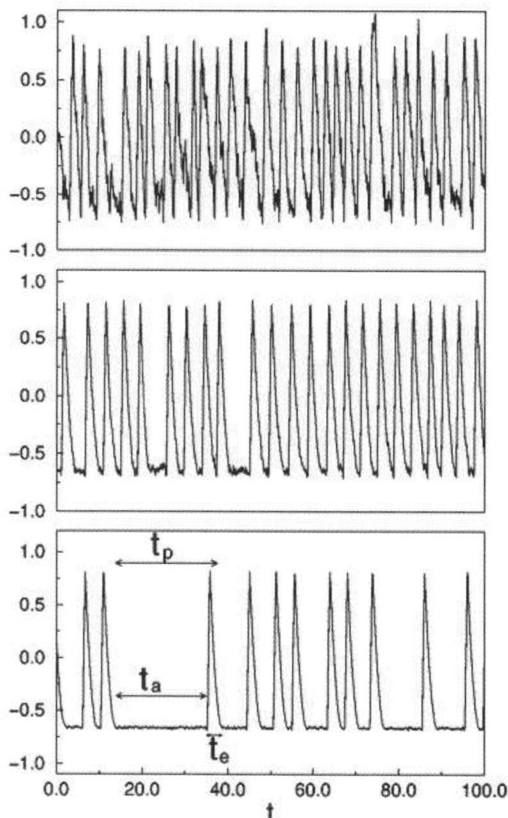

Figure 8.5: Coherence resonance. Dynamics of an excitable system perturbed by stochastic fluctuations for different values of the noise amplitude increasing from top to bottom. At intermediate noise levels, nearly periodic spikes occur (from Pikovsky and Kurths (1997)).

The effect of mixing in a spatially extended excitable medium with additive noise has been studied by Zhou et al. (2003). The external perturbation $\xi(\mathbf{r}, t)$ was assumed to be Gaussian and delta-correlated in both space and time, satisfying

$$\langle \xi(\mathbf{r}, t)\xi(\mathbf{r}', t')\rangle = 2\Gamma\delta(\mathbf{r} - \mathbf{r}')\delta(t - t').$$

The behavior of the concentration fields was studied as a function of the stirring rate ν (defined as the inverse period of the flow) and noise intensity Γ. In the absence of stirring, localized noise induced excitations develop, that propagate by diffusion and produce a random pattern of small excited patches, since short-range diffusive coupling alone is unable to generate coherent behavior in the whole extended system. With stirring, if the noise intensity is relatively small, the system does not produce any visible excitations. This is because the stirring suppresses the excitations by diluting them before they could develop and propagate in space. For somewhat larger noise amplitudes, however, an almost exactly time-periodic excitation of the whole system occurs. This takes place by the appearance of a single small localized excitation center that is strong enough to avoid dilution by the flow and very quickly propagates in space and covers the whole domain through fast advective transport. Following the almost synchronous excitation, the system returns to the steady state and the process repeats itself when a new excitation center develops, producing a cycle of regular sustained coherent oscillations of the mean field with a well defined frequency.

The frequency of the oscillations increases with the noise amplitude (Fig. 8.6). This is simply a consequence of the shorter waiting time needed for the development of a new excitation center. When the noise level is increased further, after the first coherent excitation the return to the homogeneous steady state does not happen uniformly in space. Due to the local fluctuations the coherence of the system is lost and the mean field is almost constant, while the spatial structure consists of a continuously changing mixture of randomly distributed excited filaments that do not develop a global excitation. When the stirring rate is increased, the three different regimes are shifted towards higher noise intensities.

Thus the stirred excitable media under random perturbations can produce regular coherent oscillations whose frequency is controlled by the intensity of mixing. Note that this is in contrast with the

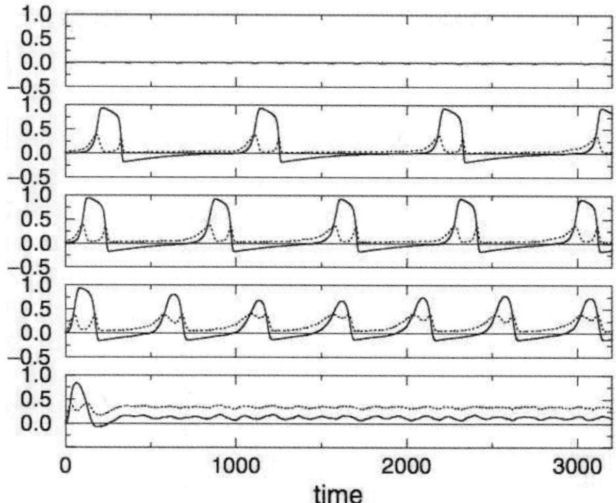

Figure 8.6: Spatially averaged concentrations (solid C_1, dashed C_2) of a noisy excitable system stirred by a chaotic flow for different noise intensities, increasing from top to bottom (from Zhou et al. (2003)).

non-uniform oscillatory media, where the stirring affects the amplitude of the oscillations of the mean field, but leaves the oscillation period unchanged. Thus, there are distinct signatures of the two mechanisms on experimentally accessible quantities.

8.4 The effect of chaotic dispersion on cyclic competition

Another example of the interplay between chaotic mixing and oscillations is the problem of cyclic competition of microorganisms dispersed in a moving fluid medium. In simple population dynamics models with strong interspecific competition (see Sect. 3.2.3), usually the species having a competitive advantage over another invades its territory leading to the extinction of the competitor. Multispecies systems, however, do not always have a simple competition hierarchy, due to the possibility of non-transitive pairwise competitive advantage of the species. This means, that the advantage of a species A

over B, and the advantage of B over a third species C, does not necessarily implies the competitive advantage of A over C. Therefore in the case of such cyclic competition, there is no obvious winner and loser in the system, somewhat similarly to the strategies in the so called rock-paper-scissors game.

Durrett and Levin (1998) considered a simple lattice model occupied by three species in cyclic competition and observed that the behavior of the system in a spatially extended system with short range local interaction is different from the corresponding mean-field model. In general, cyclic competition in spatially extended systems produces a dynamical equilibrium in which all species coexist, while the mean-field model leads to either periodically oscillating total populations, or extinction of all except one of the species.

The different behavior of the spatially distributed and of the homogeneously well mixed systems was also confirmed by the experiments of Kerr et al. (2002), who studied cyclic competition of three strains of *E. coli* bacteria in a well mixed liquid environment and on a solid agar surface. In these experiments a "killer" strain (R), with a relatively low fitness, i.e. low reproduction rate, can eliminate the "sensitive" type cells (S) by producing a toxic substance. A third "resistant" strain (R), however, is immune to the toxin and has higher fitness than the "killer" cells, but is less fit than the "sensitive" ones, due to the metabolic cost of producing proteins for protection from the toxin.

When the three strains of bacteria are grown on the agar surface they coexist, but in a continuously shaken flask two of the three strains die out after a few generations. The different outcomes in the well mixed and the static environments can also be reproduced numerically, in a simple cellular automata type model where the competing cells are distributed on a rectangular lattice. In this model each cell can die or reproduce into adjacent empty sites according to stochastic update rules. If a randomly selected lattice site is empty, it is filled with one of the cell types according to a set of probabilities, p_K, p_S, p_R, that are equal to the proportion of each cell type among the eight first neighbors. When a non-empty site is selected, the cell dies with probabilities δ_K, δ_S and δ_R. The death rates of K and R are fixed constants, while the death rate of the sensitive cells depends on the number of killer cells among the neighboring sites. Thus, $\delta_S = \delta_S^0 + \tau p_K$, where τ is a parameter characterizing the strength of the toxin. This system produces cyclic competition when the parameters satisfy the conditions: $\delta_K > \delta_R > \delta_S^0$ and $\delta_S^0 + \tau > \delta_K$.

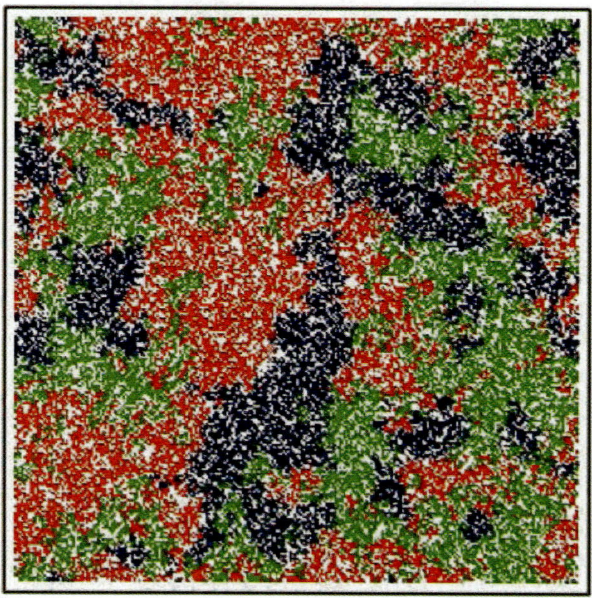

Figure 8.7: Coexistence of the "killer" (red), "sensitive" (blue) and "resistant" (green) strains in a lattice model with first-neighbor interaction (from Kerr et al. (2002)).

To connect the two markedly different scenarios observed in the static and the well-mixed environments, it is natural to analyze the role of increasing mobility (Reichenbach et al., 2007). Karolyi et al. (2005) studied the above competition model combined with dispersion by a chaotic map that represents advection of fluid elements in the alternating sine-flow. By continuously changing the frequency of the chaotic dispersion as a control parameter, it is possible to follow the transitions between the two limiting situations. When the chaotic mixing is much faster than the local population dynamics, the killer and resistant cells gradually disappear from the population and only the sensitive cells survive. This is because the killer cells

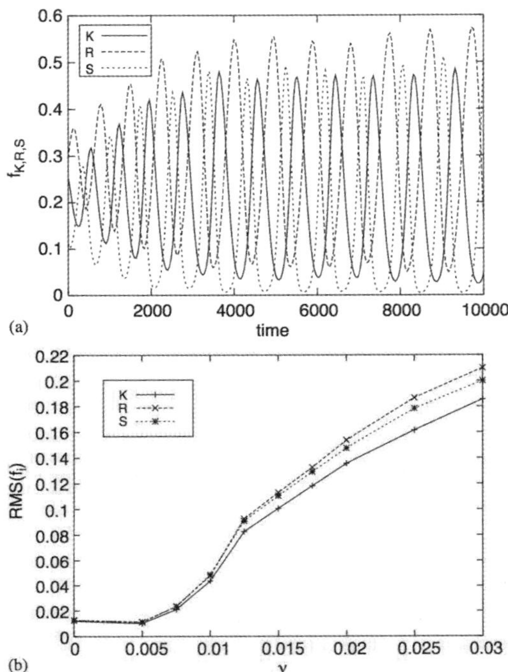

Figure 8.8: Fraction of lattice sites occupied by each species in cyclic competition below the critical stirring rate (top), and the oscillation amplitude as a function of the stirring rate ν (from Karolyi et al. (2005)).

are not efficient when strong mixing disperses and dilutes the toxin. In this limit, the behavior of the system is well approximated by the mean field model, in which the fraction of lattice sites occupied be the three species is described by the equations (compare with (3.89))

$$
\begin{aligned}
\frac{df_K}{dt} &= f_K(1 - f_S - f_R - f_K) - \delta_K f_K \\
\frac{df_R}{dt} &= f_R(1 - f_S - f_R - f_K) - \delta_R f_R \\
\frac{df_S}{dt} &= f_S(1 - f_S - f_R - f_K) - (\delta_S^0 + \tau f_K) f_S.
\end{aligned}
\tag{8.28}
$$

The first terms on the r.h.s. of these equations represent the reproduction rate of each species that is proportional to their population

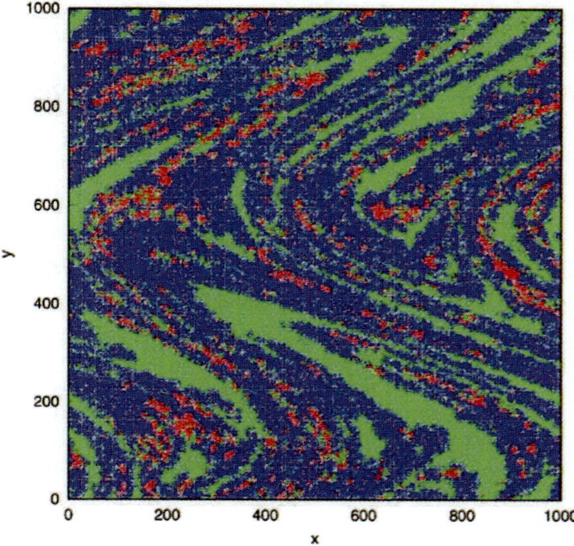

Figure 8.9: Typical distribution of the competing strains – "killer" (red), "sensitive" (blue) and "resistant" (green) – in the presence of chaotic dispersion by the sine-flow given by Eq. (2.66).

size and to the number of empty lattice sites. The system has five steady states, but given the assumptions about the parameters, the only stable steady state is $f_K = f_R = 0, f_S = 1 - \delta_S^0$, i.e. for long times the only survivor is the sensitive strain. In the other limit with only short range interaction and no dispersion, all three strains coexist in patches that continuously invade each other in a cyclic fashion (Fig. 8.7). In this case each strain maintains a roughly constant total population size.

When the chaotic dispersion is slow the populations of the three strains oscillates periodically in time, each strain becoming temporarily dominant in a cyclic fashion (Fig. 8.8). The chaotic dispersion also affects the spatial structure, by stretching the patches occupied by different strains into elongated filaments (Fig. 8.9). Thus the effect of chaotic dispersion is to synchronize the local oscillations over the whole system. The amplitude of the oscillations increases with the dispersion rate and eventually leads to the extinction of two of the three species at a critical dispersion rate. The species that dies out first depends on the parameters, being simply the one with the smallest average population size in the system with no mixing.

An order-of-magnitude estimate of the critical stirring rate can be obtained from the balance between the typical flow velocity and the speed of the propagating fronts. The assumption behind this is that the flow produces a topological transition by generating a quasi-one-dimensional filamental structure. This naturally leads to the break down of coexistence, due to the different invasion velocities of the fronts separating the patches of different strains.

Chapter 9

Further Reading

In the previous chapters we discussed various aspects of chemical and biological activity in fluid flows presenting certain classes of dynamical behavior that can be described by reaction-diffusion-advection equations and analyzed using dynamical systems approaches. However, there are many research areas of chemical or biological processes taking place in fluid environments that were not covered in the previous chapters. Here we briefly discuss some of these areas and point the reader to the relevant literature for further reading. Apart from classical well-studied topics, here we also focus on more recent developments and active areas of current research.

9.1 Complex fluids and reactive flows

The role of mixing has been studied in systems with more complex reaction schemes or considering more complex fluid-dynamical properties, and in the context of chemical engineering or microfluidic applications (for reviews on microfluidics see e.g. Squires (2005) or Ottino and Wiggins (2004)). Muzzio and Liu (1996) studied bimolecular and so-called competitive-consecutive reactions with multiple timescales in chaotic flows. Reduced models that predict the global behavior of the competitive-consecutive reaction scheme were introduced by Cox (2004) and by Vikhansky and Cox (2006), and a method for statistical description of reactive flows based on a conditional moment closure approach was given in Vikhansky and Cox (2007).

The effect of mixing in multiphase fluids (composed for example of two immiscible liquids, such as oil and vinegar), can be described by coupling the Cahn-Hilliard equation to simple model flows (Berthier et al., 2001; Naraigh and Thiffeault, 2007) or Navier-Stokes turbulence (Berti et al., 2005). It was shown that in these systems the competing effects of phase separation and advective flow can slow down or arrest the coarsening process characteristic to these systems. The dynamics of non-Newtonian viscoelastic polymer solutions was studied experimentally by Groisman and Steinberg (2001) or Arratia et al. (2006). These works have demonstrated that the elastic polymer molecules create flow instabilities that produce aperiodic time-dependent flows and efficient mixing even at very low Reynolds numbers (Groisman and Steinberg, 2000). For theoretical and numerical work on polymer solutions see e.g. Berti et al. (2006), Balkovsky et al. (2001b), or Chertkov (2000).

9.2   Self-propelled particles in prescribed flows

Fluid transport processes play an important role in the dispersal of swimming microorganisms like bacteria, algae etc. The swimming motion of motile microorganisms is often described as a random walk that for large length scales and times can be treated as a diffusion process. However, the swimming direction of the cells can be biased by various external factors, like preferential swimming towards light (phototaxis) or chemoattractants (chemotaxis), etc. In some cases the preferential swimming direction arises due to physical constraints like non-uniform mass distribution that in the case of bottom-heavy algal cells results in an upwards swimming motion (gyrotaxis). The dispersion of such gyrotactic algal cells in a vertical shear flow was studied by Hill and Bees (2002) using a Focker-Planck equation description for the probability distribution of cells in space and orientation. From this they obtained expressions for the mean transport velocity and components of the diffusion tensor. Another type of preferential swimming orientation relative to the ambient flow can arise from hydrodynamic interactions due the viscous torque exerted by the fluid on a particle, that influences its swimming direction. Torney and Neufeld (2007) considered self-propelled elongated particles in simple cellular flows and have shown that the reorientation by the flow leads to a non-uniform distribution of the particles that accumulate in hyperbolic flow regions.

The motion of chemotactic cells can be described as a biased random walk with net transport towards higher concentrations by performing a so called run-and-tumble motion, where straight runs are alternated by tumbling when a new random direction is selected for the following run. The chemotactic bias is introduced for example by making the length of the runs dependent on the relative changes in the concentration along the trajectory. The effect of shear flow on the run-and-tumble type chemotaxis of spherical cells was studied by Bearon and Pedley (2000) and Bearon (2003) who derived expressions for the dependence of the macroscopic characteristics of the chemotaxis on the strength of the shear. This has been extended to elongated cells and different types of chemotactic responses by Locsei and Pedley (2009). Torney and Neufeld (2008) considered preferential swimming in the direction of the gradient of an external field (e.g. in phototaxis) in time-dependent smooth turbulent flows. In this case the chemotactic particles aggregate on fractal filaments whose spatial structure is somewhat similar to the ones observed in the distribution of advected inertial particles. The fractal dimension of the distributions depends on the strength of the taxis and can be given analytically as a function of the non-dimensional ratio of the average kinetic energy of swimming to the energy of the turbulent carrier flow.

Considering slightly larger scales the problem of locating the source of an odor in a turbulent environment was studied by Balkovsky and Shraiman (2002). They proposed a search algorithm based on active exploration in form of counter-turning trajectories similar to the observed search behavior of moths. Using a simple statistical representation of the flow it was shown that this strategy is more efficient than standard methods based on a maximum-likelihood approach. A further extension of this search strategy, that balances the time spent on exploration and exploitation of the available information is "infotaxis" (Vergassola et al., 2007), that selects trajectories by maximizing the expected gain of information.

9.3   Bioconvection driven by swimming cells

It has been observed experimentally that swimming microorganisms in dilute suspension can generate macroscopic fluid motions that appear to be similar to flow patterns in thermal convection (see e.g. Pedley and Kessler (1992), Bees and Hill (1997), or Jánosi et al. (1998)). In the phenomenon known as bioconvection the microorganisms tend to swim upwards on average and accumulate near the free surface of the fluid. This can be caused either by non-uniform mass distribution of the cells (gyrotaxis) or due to preferential swimming towards light or higher oxygen concentration. Since the density of cells is somewhat higher than the density of water, the accumulation of cells in the upper layer produces an unstable stratification leading to overturning instability in form of sinking plumes of high cell density, as in the Rayleigh-Taylor instability. The macroscopic convective fluid motion is maintained further by the swimming motion of the microorganisms against the gravitational field.

Theoretical work on bioconvection is mainly based on the continuum description and assumes dilute suspensions in which the interactions between the cells can be neglected. Using Navier-Stokes equation for the flow and advection-diffusion equations for cell density and including the effects of cell motility, the stability analysis of the system can predict characteristic lengthscales of the convective flow and the conditions for the onset of the instability. Nonlinear behavior has been studied by numerical simulation using the continuum description (Ghorai and Hill, 2007), and alternative particle-based models of bioconvection were developed by Hopkins and Fauci (2002). For reviews on bioconvection see Pedley and Kessler (1992), and more recently Hill and Pedley (2005). A similar but different type of macroscopic flow generated by swimming cells is the so called chemotactic Boycott effect (Tuval et al., 2005). This was demonstrated experimentally in sessile drops containing oxytactic bacteria that drive a vortex flow leading to accumulation of cells near the air-water-solid contact line.

9.4 Mesoscopic flows in active suspensions

While in bioconvection macroscopic flows are generated through the gravitational field acting on the non-uniform density distribution, coherent flows can also arise in dense suspension of swimming cells directly through hydrodynamic interactions, independently of the external gravitational forces. Mesoscopic jets and vortices on scales of 50-100 microns, that is much larger than the size of individual cells, have been observed in suspensions of bacteria (Dombrowski et al., 2004; Sokolov et al., 2007). In spite of the very low Reynolds numbers characteristic at such small scales, these flows are reminiscent of turbulent flows and produce enhanced diffusivity and anomalous diffusion of advected passive particles (Wu and Libchaber, 2000). Numerical simulations have shown that self-propelled particles exerting a force dipole on the fluid can generate large scale coherent vortex motions and anomalous diffusion (Hernandez-Ortiz et al., 2005) consistently with experimental observations. Further work also studied the onset of instability in aligned or isotropic suspensions and investigated numerically the nonlinear dynamics for different types of swimming mechanisms using particle-based or continuum models (Aranson et al., 2007; Saintillan and Shelley, 2007, 2008; Ishikawa and Pedley, 2008).

The description of mixing and its effects on chemical and biological processes taking place in fluid environments has been an area of very active research over the last decade or so. While the range of problems studied covers a very wide range of complex nonlinear systems, the combination of theoretical, computational and experimental work led to significant progress in understanding the behavior of these systems. Perhaps the current challenge now is to use the newly developed insights and approaches to develop useful practical applications e.g. through the design of microfluidic devices, in biotechnology applications, in monitoring geophysical systems, etc.

Bibliography

M. Abel, A. Celani, D. Vergni, and A. Vulpiani. Front propagation in laminar flows. *Phys. Rev. E*, 64: 46307, 2001.

M. Abel, M. Cencini, D. Vergni, and A. Vulpiani. Front speed enhancement in cellular flows. *Chaos*, 12: 481, 2002.

E.R. Abraham. The generation of plankton patchiness by turbulent stirring. *Nature*, 391: 577–580, 1998.

E.R. Abraham and E.M. Bowen. Chaotic stirring by a mesoscale surface-ocean flow. *Chaos*, 12: 373, 2002.

E.R. Abraham, C.S. Law, P.W. Boyd, S.J. Lavender, M.T. Maldonado, and A.R. Bowie. Importance of stirring in the development of an iron-fertilized phytoplankton bloom. *Nature*, 407: 727, 2000.

W.C. Allee. *The social life of animals*. Norton and Co., New York, 1938.

S.M. Allen and J.W. Cahn. A microscopic theory for antiphase boundary motion and its application to antiphase domain coarsening. *Acta Metall.*, 27: 1085–1095, 1979.

D. Alonso, F. Bartomeus, and J. Catalan. Mutual interference between predators can give rise to Turing spatial patterns. *Ecology*, 83: 28, 2002.

T.M. Antonsen, Z. Fan, E. Ott, and E. Garcia-Lopez. The role of chaotic orbits in the determination of power spectra of passive scalars. *Phys. Fluids*, 8: 3094–3104, 1996.

I.S. Aranson and L. Kramer. The world of the complex Ginzburg-Landau equation. *Rev. Mod. Phys.*, 74: 99–143, 2002.

I.S. Aranson, A. Sokolov, J.O Kessler, and R.E. Goldstein. Model for dynamical coherence in thin films of self-propelled microorganisms. *Phys. Rev. E*, 75, 2007.

H. Aref. Integrable, chaotic, and turbulent vortex motion in two-dimensional flows. *Ann. Rev. Fluid Mech.*, 15: 345–389, 1983.

H. Aref. Stirring by chaotic advection. *J. Fluid Mech.*, 143: 1–21, 1984.

H. Aref. The development of chaotic advection. *Phys. Fluids*, 14: 1315–1325, 2002.

V.I. Arnold. Sur la topologie des écoulements stationnaries des fluides parfaits. *C.R. Acad. Sci. Paris*, 261: 17–21, 1965.

D.G. Aronson and H.F. Weinberger. Multidimensional nonlinear diffusion arising in population dynamics. *Adv. Math.*, 30: 33, 1978.

P.E. Arratia and J.P. Gollub. Predicting the progress of diffusively limited chemical reactions in the presence of chaotic advection. *Phys. Rev. Lett.*, 96(2): 024501, 2006.

P.E. Arratia, C.C. Thomas, J. Diorio, and J.P. Gollub. Elastic instabilities of polymer solutions in cross-channel flow. *Phys. Rev. Lett.*, 96: 144502, 2006.

V. Artale, G. Boffetta, A. Celani, M. Cencini, and A. Vulpiani. Dispersion of passive tracers in closed basins: Beyond the diffusion coefficient. *Phys. Fluids*, 9: 3162–3171, 1997.

E. Aurell, G. Boffetta, A. Crisanti, G. Paladin, and A. Vulpiani. Predictability in the large: an extension of the concept of Lyapunov exponent. *J. Phys. A*, 30: 1, 1997.

A. Babiano, J.H.E. Cartwright, O. Piro, and A. Provenzale. Dynamics of a small neutrally buoyant sphere in a fluid and targeting in hamiltonian systems. *Phys. Rev. Lett.*, 84: 5764, 2000.

E. Balkovsky, G. Falkovich, and A. Fouxon. Intermittent distribution of inertial particles in turbulent flows. *Phys. Rev. Lett.*, 86: 2790–2793, 2001a.

E. Balkovsky and A. Fouxon. Universal long-time properties of Lagrangian statistics in the Batchelor regime and their application to the passive scalar problem. *Phys. Rev. E*, 60: 4164–4174, 1999.

E. Balkovsky, A. Fouxon, and V. Lebedev. Turbulence of polymer solutions. *Phys. Rev. E*, 64: 056301, 2001b.

E. Balkovsky and B. Shraiman. Olfactory search at high Reynolds number. *Proc. Nat. Acad. Sci.*, 99: 12589, Jan 2002.

F. Bartomeus, D. Alonso, and J. Catalan. Self-organized spatial structures in a ratio-dependent predator-prey model. *Physica A*, 295: 53, 2001.

G.K. Batchelor. *The Theory of Homogeneous Turbulence*. Cambridge University Press, Cambridge, 1953.

G.K. Batchelor. Small scale variation of convected quantities like temperature in turbulent fluid. Part 1. General discussion and the case of small conductivity. *J. Fluid Mech.*, 5: 113, 1959.

G.K. Batchelor. *An Introduction to Fluid Dynamics*. Cambridge University Press, Cambridge, 1967.

G.K. Batchelor. Computation of the energy spectrum in homogeneous two-dimensional turbulence. *Phys. Fluids*, 12: 233, 1969.

R.N. Bearon. An extension of generalized Taylor dispersion in unbounded homogeneous shear flows to run-and-tumble chemotactic bacteria. *Phys. Fluids*, 15: 1552–1563, 2003.

R.N. Bearon and T.J. Pedley. Modelling run-and-tumble chemotaxis in a shear flow. *Bull. Math. Biol.*, 62: 775–791, 2000.

J. Bec. Fractal clustering of inertial particles in random flows. *Phys. Fluids*, 15: 81, 2003.

J. Bec. Multifractal concentrations of inertial particles in smooth random flows. *J. Fluid Mech.*, 528: 255–277, 2005.

M. Bees and N. Hill. Wavelengths of bioconvection patterns. *J. Exp. Biol.*, 200: 1515–1526, 1997.

B.P. Belousov. A periodic reaction and its mechanism. 1951. Original unpublished. Traslated to English in Field and Burger (1985), pp. 605-613.

I.J. Benczik, Z. Toroczkai, and T. Tél. Selective sensitivity of open chaotic flows on inertia l tracer advection: Catching particles with a stick. *Phys. Rev. Lett.*, 89: 164501, 2002.

A.F. Bennett. Relative dispersion: Local and nonlocal dynamics. *J. Atmosph. Sci.*, 41: 1881–1886, 1984.

L. Berthier, J.-L. Barrat, and J. Kurchan. Phase separation in a chaotic flow. *Phys. Rev. Lett.*, 86: 2014–2017, 2001.

S. Berti, A. Bistagnino, G. Boffetta, A. Celani, and S. Musacchio. Small-scale statistics of viscoelastic turbulence. *Europhys. Lett.*, 76: 63–69, 2006.

S. Berti, G. Boffetta, M. Cencini, and A. Vulpiani. Turbulence and coarsening in active and passive binary mixtures. *Phys. Rev. Lett.*, 95: 224501, 2005.

D.A. Birch, Y.-K. Tsang, and W. R. Young. Bounding biomass in the Fisher equation. *Phys. Rev. E*, 75: 066304, 2007.

D.A. Birch, W. R. Young, and P. J.S. Franks. Thin layers of plankton: Formation by shear and death by diffusion. *Deep-Sea Res. I*, 55: 277 – 295, 2008.

E.S. Bish and W.J.A. Dahm. Strained dissipation and reaction layer analyses of nonequilibrium chemistry in turbulent reacting flows. *Combust. Flame*, 100: 457–464, 1995.

W.J. Blake, M. Kaern, Ch. R. Cantor, and J. J. Collins. Noise in eukaryotic gene expression. *Nature*, 422: 633–637, 2003.

G. Boffetta, A. Celani, S. Musacchio, and M. Vergassola. Intermittency in two-dimensional Ekman-Navier-Stokes turbulence. *Phys. Rev. E*, 66: 026304, 2002.

G. Boffetta, M. Cencini, S. Espa, and G. Querzoli. Chaotic advection and relative dispersion in an experimental convective flow. *Phys. Fluids*, 12: 3160–3167, 2000.

G. Boffetta, A. Cenedese, S. Espa, and S. Musacchio. Effects of friction on 2d turbulence: An experimental study of the direct cascade. *Europhys. Lett.*, 71: 590–596, 2005.

G. Boffetta, G. Lacorata, G. Radaelli, and A. Vulpiani. Detecting barriers to transport: a review of different techniques. *Physica D*, 159: 58–70, 2001.

T. Bohr, M. Jensen, G. Paladin, and A. Vulpiani. *Dynamical Systems Approach to Turbulence*. Cambridge Univ. Press, Cambridge, 1998.

A. Bray. Theory of phase-ordering kinetics. *Advances in Physics*, 43: 357–459, 1994. Reprinted in Adv. Phys. 51, 481–587 (2002).

G.E. Briggs and J.B.S. Haldane. A note on the kinetics of enzyme action. *Biochem. J.*, 19: 338–339, 1925.

P. Camacho and J.D. Lechleiter. Increased frequency of calcium waves in *Xenopus laevis* oocites that express a calcium-ATPase. *Science*, 260: 226–229, 1993.

G.F. Carrier, F.E. Fendell, and F.E. Marble. The effect of strain rate on diffusion flames. *SIAM J. Appl. Math.*, 28: 463–500, 1975.

J.H.E. Cartwright, M. Feingold, and O. Piro. Global diffusion in a realistic three-dimensional time-dependent nonturbulent fluid flow. *Phys. Rev. Lett.*, 75: 3669–72, 1995.

J.H.E. Cartwright, M. Feingold, and O. Piro. Chaotic advection in three-dimensional unsteady incompressible laminar flow. *J. Fluid Mech.*, 316: 259–284, 1996.

V. Castets, E. Dulos, J. Boissonade, and P. De Kepper. Experimental evidence of a sustained Turing-type nonequilibrium chemical pattern. *Phys. Rev. Lett.*, 64: 2953–2956, 1990.

J. Catalan. Small-scale hydrodynamics as a framework for plankton evolution. *Jpn. J. Limnol.*, 60: 469–494, 1999.

H.J. Catrakis, R.C. Aguirre, and J. Ruiz-Plancarte. Area and volume properties of fuid interfaces in turbulence: scale-local self-similarity and cumulative scale dependence. *J. Fluid Mech.*, 462: 245, 2002.

M. Cencini, C. López, and D. Vergni. Reaction-diffusion systems: front propagation and spatial structures. In A. Vulpiani and R. Livi, editors, *The Kolmogorov Legacy in Physics*. Springer-Verlag, 2003a.

M. Cencini, A. Torcini, D. Vergni, and A. Vulpiani. Thin front propagation in steady and unsteady cellular flows. *Phys. Fluids*, 15: 679, 2003b.

J. Chaiken, R. Chevray, M. Tabor, and Q.M. Tan. Experimental study of Lagrangian turbulence in a Stokes flow. *Proc. Roy. Soc. London A*, 408: 165–174, 1986.

R.N. Chapman. *Animal Ecology*. McGraw-Hill, New York, 1931.

H. Chaté. Spatiotemporal intermittency regimes of the one-dimensional complex ginzburg-landau equation. *Nonlinearity*, 7: 185–204, 1994.

H. Chaté. Disordered regimes of the one-dimensional complex Ginzburg-Landau equation. In P. Cladis and P. Palffy-Muhoray, editors, *Spatiotemporal Patterns in Nonequilibrium Complex Systems*, Vol. XXI of Santa Fe Institute in the Sciences of Complexity, pages pp. 5–49. Addison-Wesley, New York, 1995.

H. Chaté and P. Manneville. Phase diagram of the two-dimensional complex Ginzburg-Landau equation. *Physica A*, 224: 348–368, 1996.

M. Chertkov. On how a joint interaction of two innocent partners (smooth advection and linear damping) produces a strong intermittency. *Phys. Fluids*, 10: 3017–3019, 1998.

M. Chertkov. Passive advection in nonlinear medium. *Phys. Fluids*, 11: 2257–2262, 1999.

M. Chertkov. Polymer stretching by turbulence. *Phys. Rev. Lett.*, 84: 4761–4764, 2000.

M. Chertkov and V. Lebedev. Boundary effects on chaotic advection-diffusion chemical reactions. *Phys. Rev. Lett.*, 90: 134501, 2003.

S. Corrsin. On the spectrum of isotropic temperature field in isotropic turbulence. *J. Appl. Phys.*, 22: 469–473, 1951.

S. Corrsin. The reactant concentration spectrum in turbulent mixing with a first-order reaction. *J. Fluid Mech.*, 11: 407–416, 1961.

S.M. Cox. Chaotic mixing of a competitive-consecutive reaction. *Physica D*, 199: 369–386, 2004.

S.M. Cox and G.A. Gottwald. A bistable reaction-diffusion system in a stretching flow. *Physica D*, 216: 307–318, 2006.

M.C. Cross and P. C. Hohenberg. Pattern formation outside of equilibrium. *Rev. Mod. Phys.*, 65(3): 851–1112, Jul 1993.

S. Dano, P.G. Sorensen, and F. Hynne. Sustained oscillations in living cells. *Nature*, 402: 320, 1999.

J.M. Davidenko, A.V. Pertsov, R. Salomonsz, W. Baxter, and J. Jalife. Stationary and drifting spiral waves of excitation in isolated cardiac muscle. *Nature*, 355: 349–351, 1992.

B. Dennis. Allee effects: population growth, critical density, and the chance of extinction. *Nat. Res. Model.*, 3: 481–538, 1989.

T. Dombre, U. Frisch, J.M. Greene, M. Hénon, A. Mehr, and A.M. Soward. Chaotic streamlines in the ABC flows. *J. Fluid Mech.*, 167: 353–391, 1986.

C. Dombrowski, L. Cisneros, S. Chatkaew, R.E. Goldstein, and J.O. Kessler. Self-concentration and large-scale coherence in bacterial dynamics. *Phys. Rev. Lett.*, 93: 098103, 2004.

F. d'Ovidio, V. Fernández, E. Hernández-García, and C. López. Mixing structures in the Mediterranean Sea from Finite-Size Lyapunov exponents. *Geophys. Res. Lett.*, 31: L17203, 2004.

F. d'Ovidio, J. Isern-Fontanet, C. López, E. Hernández-García, and E. García-Ladona. Comparison between Eulerian diagnostics and finite-size Lyapunov exponents computed from altimetry in the Algerian basin. *Deep-Sea Res. I*, 56(1): 15 – 31, 2009.

D.M. Dubois. A model of patchiness for prey-predator plankton populations. *Ecol. Mod.*, 1: 67–80, 1975a.

D.M. Dubois. Simulation of the spatial structuration of a patch of prey-predator plankton populations in the Southern Bight of the North Sea. *Mém. Soc. Roy. Sci. Liège Ser. 6*, 7: 75–82, 1975b.

R. Durrett. *Stochastic Calculus: A practical introduction.* CRC-Press, Cornell University, Ithaca, New York, USA, 1996.

R. Durrett and S. Levin. Spatial aspects of interspecific competition. *Theor. Pop. Biol.*, 53: 30–43, 1998.

U. Ebert and W. van Saarloos. Front propagation into unstable states: Universal algebraic convergence towards uniformly translating pulled fronts. *Physica D*, 146: 1–99, 2000.

J.P. Eckmann and D. Ruelle. Ergodic theory of chaos and strange attractors. *Rev. Mod. Phys.*, 57: 617–656, 1985. Addenda in Rev. Mod. Phys. 57, 1115 (1985).

S. Edouard, B. Legras, F. Lefèvre, and R. Aymard. The effect of small-scale inhomogeneities on ozone depletion in the Artic. *Nature*, 384: 444, 1996.

M. Edwards and A. Yool. The role of higher predation in plankton population models. *J. Plank. Res.*, 22: 1085–1112, 2000.

I.R. Epstein. The consequences of imperfect mixing in autocatalytic chemical and biological systems. *Nature*, 374: 321–327, 1995.

I.R. Epstein, C.E. Dateo, P. De Kepper, K. Kustin, and M. Orbán. Bistability in a C.S.T.R.: New experimental examples and mathematical modeling. In C. Vidal and A. Pacault, editors, *Nonlinear Phenomena in Chemical Dynamics*, pages 188–191. Springer Verlag, 1981.

I.R. Epstein and J. A. Pojman. *An Introduction to Nonlinear Chemical Dynamics: Oscillations, Waves, Patterns, and Chaos.* Oxford University Press, Oxford, 1998.

G. Falkovich, K. Gawędzki, and M. Vergassola. Particles and fields in fluid turbulence. *Rev. Mod. Phys.*, 73: 913–975, 2001.

M. Farkas. *Dynamical Models in Biology.* Academic Press, San Diego, 2001.

M. Feingold, L.P. Kadanoff, and O. Piro. Passive scalars, three-dimensional volume preserving maps, and chaos. *J. Stat. Phys.*, 50: 529–565, 1988.

D.R. Fereday and P.H. Haynes. Scalar decay in two-dimensional chaotic advection and Batchelor-regime turbulence. *Phys. Fluids*, 16: 4359, 2004.

D.R. Fereday, P.H. Haynes, A. Wonhas, and J.C. Vassilicos. Scalar variance decay in chaotic advection and Batchelor-regime turbulence. *Phys. Rev. E*, 65: 035301, 2002.

R.J. Field and M. Burger, editors. *Oscillations and Traveling Waves in Chemical Systems.* Wiley, New York, 1985.

R.J. Field, E. Körös, and R.M. Noyes. Oscillations in chemical systems, Part 2. Thorough analysis of temporal oscillations in the bromate-cerium-malonic acid system. *J. Am. Chem. Soc.*, 94: 8649–8664, 1972.

R.J. Field and R.M. Noyes. Oscillations in chemical systems, Part 4. Limit cycle behaviour in a model of a real chemical reaction. *J. Chem. Phys.*, 60: 1877–1844, 1974.

P. Fife. Propagator-controller systems and chemical patterns. In C. Vidal and A. Pacault, editors, *Non-equilibrium dynamics in chemical systems*, pages 76–88. Springer-Verlag, Berlin, 1984.

R.A. Fisher. The wave of advance of adventageous genes. *Ann. Eugenics*, 7: 353–369, 1937.

R. FitzHugh. Impulses and physiological states in theoretical models of nerve membrane. *Biophys. J.*, 1: 445–466, 1961.

G.O. Fountain, D.V. Khakhar, and J.M. Ottino. Visualization of three-dimensional chaos. *Science*, 281: 683–686, 1998.

U. Frisch. *Turbulence.* Cambridge University Press, Cambridge, 1995.

C.W. Gardiner. *Handbook of Stochastic Methods for Physics, Chemistry and the Natural Sciences, 3rd Ed.* Springer-Verlag, Berlin, 2004.

S. Ghorai and N.A. Hill. Gyrotactic bioconvection in three dimensions. *Phys. Fluids*, 19: 054107, 2007.

D.T. Gillespie. Exact stochastic simulation of coupled chemical reactions. *J. Phys. Chem.*, 81: 23402361, 1977.

M. Giona, A. Adrover, and S. Cerbelli. On the use of the pulsed-convection approach for modelling advection-diffusion in chaotic flows - A prototypical example and direct numerical simulations. *Physica A*, 349: 37–73, 2005.

M. Giona, A. Adrover, S. Cerbelli, and V. Vitacolonna. Spectral properties and transport mechanisms of partially chaotic bounded flows in the presence of diffusion. *Phys. Rev. Lett.*, 92: 114101, 2004.

A. Goldbeter. *Biochemical Oscillations and Cellular Rhythms.* Cambridge University Press, 1995.

D. Gomila, P. Colet, M. San Miguel, and Oppo G.-L. Domain wall dynamics: Growth laws, localized structures and stable droplets. *Eur. Phys. J. - Special Topics*, 146: 71–86, 2007.

A. Groisman and V. Steinberg. Elastic turbulence in a polymer solution flow. *Nature*, 405: 53–55, 2000.

A. Groisman and V. Steinberg. Efficient mixing at low Reynolds numbers using polymer additives. *Nature*, 410: 905–908, 2001.

G. Hardin. The competitive exclusion principle. *Science*, 131: 1292–1297, 1960.

P.H. Haynes and J. Vanneste. What controls the decay of passive scalars in smooth flows? *Phys. Fluids*, 17: 097103, 2005.

E. Hernández-García and C. López. Sustained plankton blooms under open chaotic flows. *Ecol. Complex.*, 1: 253–259, 2004.

E. Hernández-García, C. López, and Z. Neufeld. Small-scale structure of nonlinearly interacting species advected by chaotic flows. *Chaos*, 12: 470–480, 2002.

E. Hernández-García, C. López, and Z. Neufeld. Filament bifurcations in a one-dimensional model of reacting excitable fluid flow. *Physica A*, 327: 59–64, 2003.

J.P. Hernandez-Ortiz, Ch. G. Stoltz, and M. D. Graham. Transport and collective dynamics in suspensions of confined swimming particles. *Phys. Rev. Lett.*, 95: 10.1021/ja00780a001, 2005.

N. Hill and M.A. Bees. Taylor dispersion of gyrotactic swimming micro-organisms in a linear flow. *Phys. Fluids*, 14: 2598–2605, 2002.

N.A. Hill and T.J. Pedley. Bioconvection. *Fluid Dyn. Res.*, 37: 1–20, 2005.

A.L. Hodgkin and A.F. Huxley. A quantitative description of membrane current and its application to conduction and excitation in nerve. *J. Physiol. (London)*, 117: 500–544, 1952.

J. Hofbauer and K. Sigmund. *The Theory of Evolution and Dynamical Systems*. Cambridge University Press, Cambridge, 1988.

C.S. Holling. The components of predation as revealed by a study of small-mammal predation on the european pine sawfly. *Canad. Entomol.*, 91: 293–332, 1959a.

C.S. Holling. Some characteristics of simple types of predation and parasitism. *Canad. Entomol.*, 91: 385–398, 1959b.

M.M. Hopkins and L.J. Fauci. A computational model of the collective fluid dynamics of motile micro-organisms. *J. Fluid Mech.*, 455: 149–174, 2002.

J. Huisman and F.J. Weissing. Biodiversity of plankton by species oscillations and chaos. *Nature*, 402: 407–410, 1999.

B.R. Hunt. The Hausdorff dimension of graphs of Weierstrass functions. *Proc. Am. Math. Soc.*, 126: 791, 1998.

G.E. Hutchinson. The paradox of the plankton. *Am. Nat.*, 95: 137–145, 1961.

T. Ishikawa and T. J. Pedley. Coherent structures in monolayers of swimming particles. *Phys. Rev. Lett.*, 100: 088103, 2008.

V.S. Ivlev. *Experimental ecology of the feeding of fishes.* Yale University Press, New York, 1961. Translation of the Russian original published in 1955.

E.A. Jackson. *Perspectives of nonlinear dynamics.* Cambridge Univ. Press, 1989.

I.M. Jánosi, J.O. Kessler, and V.K. Horváth. Onset of bioconvection in suspensions of *Bacillus subtilis*. *Phys. Rev. E*, 58: 4793–4800, 1998.

J. Jimenez and C. Martel. Fractal interfaces and product generation in the two-dimensional mixing layer. *Phys. Fluids A*, 3: 1261–1268, 1991.

M.C. Jullien, P. Castiglione, and P. Tabeling. Experimental observation of Batchelor dispersion of passive tracers. *Phys. Rev. Lett.*, 85: 3636, 2000.

C. Jung, T. Tél, and E. Ziemniak. Application of scattering chaos to particle transport in a hydrodynamical flow. *Chaos*, 3: 555–568, 1993.

G. Karolyi, Z. Neufeld, and I. Scheuring. Rock-scissors-paper game in a chaotic flow: The effect of dispersion on the cyclic competition of microorganisms. *J. Theor. Biol.*, 236: 12, 2005.

G. Károlyi, Á. Péntek, I. Scheuring, T. Tél, and Z. Toroczkai. Chaotic flow: the physics of species coexistence. *Proc. Natl. Acad. Sci.*, 97: 13661–13665, 2000.

G. Károlyi, Á Péntek, Z. Toroczkai, T. Tél, and C. Grebogi. Chemical or biological activity in open flows. *Phys. Rev. E*, 59: 5468–5481, 1999.

G. Karolyi and T. Tel. Chaotic tracer scattering and fractal basin boundaries in a blinking vortex-sink system. *Phys. Rep.*, 290: 125, 1997.

G. Károlyi and T. Tél. Chemical transients in closed chaotic flows: The role of effective dimensions. *Phys. Rev. Lett.*, 95: 264501, 2005.

G. Károlyi and T. Tél. Effective dimensions and chemical reactions in fluid flows. *Phys. Rev. E*, 76: 046315, 2007.

J. Keener and J. Sneyd. *Mathematical Physiology*. Springer-Verlag, New York, 1998.

H. Kellay, X.-I. Wu, and W.I. Goldburg. Experiments with turbulent soap films. *Phys. Rev. Lett.*, 74: 3975, 1995.

B. Kerr, M. A. Riley, M. W. Feldman, and B. J. M. Bohannan. Local dispersal promotes biodiversity in a real-life game of rock-paper-scissors. *Nature*, 418: 171–174, 2002.

H. Kierstead and J.B. Slobodkin. The size of water masses containing plankton blooms. *J. Mar. Res.*, 12: 141–147, 1953.

I.Z. Kiss, J.H. Merkin, and Z. Neufeld. Combustion initiation and extinction in a 2d chaotic flow. *Physica D*, 183: 175–189, 2003a.

I.Z. Kiss, J.H. Merkin, and Z. Neufeld. Homogenization induced by chaotic mixing and diffusion in an oscillatory chemical reaction. *Phys. Rev. E*, 70: 026216, 2004.

I.Z. Kiss, J.H. Merkin, S.K. Scott, P.L. Simon, S. Kalliadasis, and Z. Neufeld. The structure of flame filaments in chaotic flows. *Physica D*, 176: 67–81, 2003b.

R. Klages, G. Radons, and I.M. Sokolov, editors. *Anomalous transport: Foundations and applications.* Wiley-VCH, Weinheim, 2008.

A.N. Kolmogorov. The local structure of turbulence in incompressible viscous fluid for very large Reynolds number. *Dokl. Akad. Nauk SSSR*, 30: 299–303, 1941a.

A.N. Kolmogorov. Dissipation of energy in locally isotropic turbulence. *Dokl. Akad. Nauk SSSR*, 32: 16–18, 1941b.

A.N. Kolmogorov, I. Petrovskii, and N. Piskunov. Étude de l'equation de la diffusion avec croissance de la quantité de matière et son aplication à un problème biologique (Study of the diffusion equation with growth of the quantity of matter and its application to a biology problem). *Bulletin de l'Université d'État à Moscou, Ser. Int., Sect. A (Bull. Moskv. Gos. Univ. Mat. Mekh.)*, 1: 1–25, 1937. Translated to English in Pelcé (1988) and in Tikhomirov (1991).

D.K. Kondepudi, R.J. Kaufman, and N. Singh. Chiral symmetry breaking in sodium chlorate crystallization. *Science*, 250: 975–976, 1990.

R.H. Kraichnan. Inertial ranges in 2 dimensional turbulence. *Phys. Fluids*, 10: 1417, 1967.

R.H. Kraichnan and D. Montgomery. Two-dimensional turbulence. *Rep. Prog. Phys.*, 43: 547–619, 1980.

M.J. Kreuzer. *Nonequilibrium thermodynamics and its statistical foundations.* Clarendon Press, Oxford, 1981.

Y. Kuramoto. *Chemical Oscillations, Waves, and Turbulence.* Springer-Verlag, Berlin, 1984.

G. Lacorata, E. Aurell, and A. Vulpiani. Drifter dispersion in the Adriatic Sea: Lagrangian data and chaotic model. *Ann. Geophys.*, 19: 121–129, 2001.

H. Lamb. *Hydrodynamics.* Dover, New York, 1932.

L.D. Landau and E.M. Lifschitz. *Fluid Mechanics.* Pergamon Press, Oxford, 1959.

G. Lapeyre. Characterization of finite-time Lyapunov exponents and vectors in two-dimensional turbulence. *Chaos*, 12: 688–698, 2002.

Y.-T. Lau, J.M. Finn, and E. Ott. Fractal dimension in nonhyperbolic chaotic scattering. *Phys. Rev. Lett.*, 66: 978, 1991.

M. Lesieur. *Turbulence in Fluids.* Kluwer, Dordrecht, 1990.

S.A. Levin and L.A. Segel. Hypothesis for origin of plankton patchiness. *Nature*, 259: 659, 1976.

W. Liu and G. Haller. Strange eigenmodes and decay of variance in the mixing of diffusive tracers. *Physica D*, 188: 1–39, 2004.

J. Locsei and T.J. Pedley. Run and tumble chemotaxis in a shear flow: The effect of temporal comparisons, persistence, rotational diffusion, and cell shape. *Bull. Math. Biol.*, 2009. (In press).

C. López and E. Hernández-García. The role of diffusion in the chaotic advection of a passive scalar with finite lifetime. *Eur. Phys. J. B*, 28: 353, 2002.

A.J. Lotka. Undamped oscillations derived from the law of mass action. *J. Am. Chem. Soc.*, 42: 1595–1599, 1920.

A.J. Lotka. The growth of mixed populations: two species competing for a common food supply. *J. Wash. Acad. Sci.*, 22: 461–469, 1932.

Y. Luo and I.R. Epstein. Stirring and premixing effects in the oscillatory chloriteiodide reaction. *J. Chem. Phys.*, 85: 5733–5740, 1986.

A. Mahadevan and D. Archer. Modeling the impact of fronts and mesoscale circulation on the nutrient supply and biogeochemistry of the upper ocean. *J. Geophys. Res. C*, 105: 1209–1225, 2000.

A.J. Majda and P.R. Kramer. Simplified models for turbulent diffusion: Theory, numerical modelling, and physical phenomena. *Phys. Rep.*, 314: 237–574, 1999.

K.H. Mann and J.R.N. Lazier. *Dynamics of marine ecosystems. Biological-physical interactions in the oceans.* Blackwell Scientific Publications, Boston, 1991.

A.P. Martin. On filament width in oceanic plankton distributions. *J. Plank. Res.*, 22: 597, 2000.

A.P. Martin. Phytoplankton patchiness: the role of lateral stirring and mixing. *Progr. Oceanogr.*, 57: 125–174, 2003.

L. Matthews and J. Brindley. Patchiness in plankton populations. *Dyn. Stab. Sys.*, 12: 39, 1997.

M. Maxey and J. Riley. Equation of motion of a small rigid sphere in a nonuniform flow. *Phys. Fluids*, 26: 883–889, 1983.

R.M. May. Thresholds and breakpoints in ecosystems with a multiplicity of stable states. *Nature*, 269: 471–477, 1977.

R.M. May and W. Leonard. Nonlinear aspects of competition between three species. *SIAM J. Appl. Math.*, 29: 243–252, 1975.

H.H. McAdams and A. Arkin. It's a noisy business! genetic regulation at the nanomolar scale. *Trends Genet.*, 15: 65–69, 1999.

R. McGehee and R.A. Armstrong. Some mathematical problems concerning the ecological principle of competitive exclusion. *J. Diff. Equations*, 23: 30–52, 1977.

W. McKiver and Z. Neufeld. Influence of turbulent advection on a phytoplankton ecosystem with non-uniform carrying capacity. *Phys. Rev. E*, 79: 061902, 2009.

P. McLeod, A.P. Martin, and K.J. Richards. Minimum length scale for growth-limited oceanic plankton distributions. *Ecol. Mod.*, 158: 111–120, 2002.

S.M. Menon and G.A. Gottwald. Bifurcations in reaction-diffusion systems in chaotic flows. *Phys. Rev. E*, 71: 066201, 2005.

S.N. Menon and G.A. Gottwald. Bifurcations of flame filaments in chaotically mixed reactions. *Phys. Rev. E*, 75: 16209, 2007.

M. Menzinger and A.K. Dutt. On the myth of the well-stirred reactor. the chlorite/iodide reaction. *J. Phys. Chem.*, 94: 4510, 1990.

E. Meron. Pattern formation in excitable media. *Phys. Rep.*, 218: 1–66, 1992.

G. Metcalfe and J.M. Ottino. Autocatalytic processes in mixing flows. *Phys. Rev. Lett.*, 72: 2875–2578, 1994. Erratum: Phys. Rev. Lett. 73, 212 (1994).

R. Metzler and J. Klafter. The random walks guide to anomalous diffusion: a fractional dynamics approach. *Phys. Rep.*, 339: 1–77, 2000.

M. Michaelis and M.I. Menten. Die kinetik der invertinwirkung. *Biochem. Z.*, 49: 333–369, 1913.

R.E. Mirollo and S.H. Strogatz. Amplitude death in an array of limit-cycle oscillators. *J. Stat. Phys.*, 60: 245–262, 1990.

J. Monod. *Recherches sur la Croissance des Cultures Bactériennes.* Hermann, Paris, 1942.

J.D. Murray. *Mathematical biology.* Springer-Verlag, Berlin, 1993.

F.J. Muzzio and M. Liu. Chemical reactions in chaotic flows. *Chem. Eng. J.*, 64: 117–127, 1996.

J.S. Nagumo, S. Arimoto, and S. Yoshizawa. An active pulse transmission line simulating nerve axon. *Proc. IRE*, 50: 2061–2071, 1962.

I. Nagypál and I.R. Epstein. Fluctuations and stirring rate effects in the chlorite-thiosulfate reaction. *J. Phys. Chem.*, 90: 6285–6292, 1986.

K. Nam, T.M. Antonsen, P.N. Guzdar, and E. Ott. k spectrum of finite lifetime passive scalars in Lagrangian chaotic fluid flows. *Phys. Rev. Lett.*, 83: 3426–3429, 1999.

K. Nam, E. Ott, T.M. Antonsen, and P.N. Guzdar. Lagrangian chaos and the effect of drag on the enstrophy cascade in two-dimensional turbulence. *Phys. Rev. Lett.*, 84: 5134–5137, 2000.

L. O. Naraigh and J.-L. Thiffeault. Bubbles and filaments: Stirring a Cahn-Hilliard fluid. *Phys. Rev. E*, 75: 016216, 2007.

Z. Neufeld. Excitable media in a chaotic flow. *Phys. Rev. Lett.*, 87: 108301, 2001.

Z. Neufeld, P.H. Haynes, V. Garçon, and J. Sudre. Ocean fertilization experiments may initiate a large scale phytoplankton bloom. *Geophys. Res. Lett.*, 29: 10.1029/2001GL013677, 2002a.

Z. Neufeld, P.H. Haynes, and T. Tél. Chaotic mixing induced transitions in reaction-diffusion systems. *Chaos*, 12: 426–438, 2002b.

Z. Neufeld, I.Z. Kiss, C. Zhou, and J. Kurths. Synchronization and oscillator death in oscillatory media with stirring. *Phys. Rev. Lett.*, 91: 084101, 2003.

Z. Neufeld, C. López, and P.H. Haynes. Smooth-filamental transition of active tracer fields stirred by chaotic advection. *Phys. Rev. Lett.*, 82: 2606, 1999.

Z. Neufeld, C. López, E. Hernández-García, and O. Piro. Excitable media in open and closed chaotic flows. *Phys. Rev. E*, 66: 066208, 2002c.

Z. Neufeld, C. López, E. Hernández-García, and T. Tél. The multifractal structure of chaotically advected chemical fields. *Phys. Rev. E*, 61: 3857, 2000.

P.C. Newell. Attraction and adhesion in the slime mold *dictyostelium*. In *Fungal differentiation: A contemporary synthesis*, pages pp. 43–71. Marcel Dekker, New York, 1983.

Z. Noszticzius, Zs. Bodnar, L. Garamszegi, and M. Wittmann. Hydrodynamic turbulence and diffusion-controlled reactions. simulation of the effect of stirring on the oscillating Belousov-Zhabotinsky reaction with the Radicalator model. *J. Phys. Chem.*, 95: 6575–6580, 1991.

C.R. Nugent, W.M. Quarles, and T.H. Solomon. Experimental studies of pattern formation in a reaction-advection-diffusion system. *Phys. Rev. Lett.*, 93: 218301, 2004.

A.M. Obukhov. The structure of the temperature field in a turbulent flow. *Izv. Akad. Nauk. SSSR, Ser. Geogr. and Geophys.*, 13: 469–473, 1949.

A. Okubo. Oceanic diffusion diagrams. *Deep-Sea Res.*, 18: 789, 1971.

A. Okubo and S.A. Levin. *Diffusion and Ecological Problems: Modern Perspectives*. Springer-Verlag, New York, 2002.

M. Orban, Ch. Dateo, P. De Kepper, and I. R. Epstein. Systematic design of chemical oscillators. 11. Chlorite oscillators: new experimental examples, tristability, and preliminary classification. *J. Am. Chem. Soc.*, 104: 5911–5918, 1982.

E. Ott. *Chaos in Dynamical Systems*. Cambridge Univ. Press, Cambridge (UK), 1993.

E. Ott and T. Tél. Chaotic scattering: an introduction. *Chaos*, 3: 417–426, 1993.

J.M. Ottino. *The kinematics of mixing: stretching, chaos, and transport*. Cambridge Univ. Press, Cambridge (UK), 1989a.

J.M. Ottino. The mixing of fluids. *Sci. Am.*, January 1989: 56, 1989b.

J.M. Ottino. Mixing, chaotic advection, and turbulence. *Ann. Rev. Fluid Mech.*, 22: 207–254, 1990.

J.M. Ottino. Mixing and chemical reactions: A tutorial. *Chem. Eng. Sci.*, 49: 4005–4027, 1994.

J.M. Ottino and S. Wiggins. Introduction: mixing in microfluidics. *Phil. Trans. Roy. Soc. London. A*, 362: 923–935, 2004.

Q. Ouyang and H.L. Swinney. Transition from a uniform state to hexagonal and striped Turing patterns. *Nature*, 352: 610 – 612, 1991.

O. Paireau and P. Tabeling. Enhancement of the reactivity by chaotic mixing. *Phys. Rev. E*, 56: 2287–2290, 1997.

M. Paoletti and T. Solomon. Experimental studies of front propagation and mode-locking in an advection-reaction-diffusion system. *Europhys. Lett.*, 69: 819, 2005a.

M. Paoletti and T. Solomon. Front propagation and mode-locking in an advection-reaction-diffusion system. *Phys. Rev. E*, 72: 046204, 2005b.

M.S. Paoletti, C.R. Nugent, and T.H. Solomon. Synchronization of oscillating reactions in an extended fluid system. *Phys. Rev. Lett.*, 93: 124101, 2006.

C. Pasquero, A. Bracco, and A. Provenzale. Impact of spatiotemporal variability of the nutrient flux on primary productivity in the ocean. *J. Geophys. Res.*, 110: C07005, 2005.

T.J. Pedley and J.O. Kessler. Hydrodynamic phenomena in suspensions of swimming microorganisms. *Ann. Rev. Fluid Mech.*, 24: 313–358, 1992.

P. Pelcé. *Dynamics of Curved Fronts*. Academic Press, San Diego, 1988.

V. Pérez-Muñuzuri. Resonant pattern formation in active media driven by time-dependent flows. *Phys. Rev. E*, 73: 066213, 2006.

F. Peters and C. Marrasé. Effects of turbulence on plankton: an overview of experimental evidence and some theoretical considerations. *Mar. Ecol. Prog. Ser.*, 205: 291–306, 2000.

R.T. Pierrehumbert. On tracer microstructure in the large-eddy dominated regime. *Chaos, Solitons and Fractals*, 4: 1091–1110, 1994.

R.T. Pierrehumbert. Lattice models of advection-diffusion. *Chaos*, 10: 61–74, 2000.

S. Pigolotti, C. López, and E. Hernández-García. Species clustering in competitive Lotka-Volterra models. *Phys. Rev. Lett.*, 98: 258101(1–4), 2007.

A. Pikovsky and O. Popovych. Persistent patterns in deterministic mixing flows. *Europhys. Lett.*, 61: 625–631, 2003.

A. Pikovsky, M. Rosenblum, and J. Kurths. *Synchronization: A Universal Concept in Nonlinear Sciences*. Cambridge University Press, 2003.

A.S. Pikovsky and J. Kurths. Coherence resonance in a noise-driven excitable system. *Phys. Rev. Lett.*, 78: 775, 1997.

S.B. Pope. *Turbulent Flows*. Cambridge University Press, Cambridge, 2000.

D. Poppe and H. Lustfeld. Nonlinearities in the gas phase chemistry of the troposphere: Oscillating concentrations in a simplified mechanism. *J. Geophys. Res.*, 101: 14373, 1996.

T.M. Powell and A. Okubo. Turbulence, diffusion and patchiness in the sea. *Phil. Trans. R. Soc. Lond. B*, 343: 11–18, 1994.

D. V. Ramana Reddy, A. Sen, and G. L. Johnston. Time delay induced death in coupled limit cycle oscillators. *Phys. Rev. Lett.*, 80: 5109–5112, 1998.

W.E. Ranz. Applications of a stretch model to mixing, diffusion, and reaction in laminar and turbulent flows. *AIChE J.*, 25: 41–47, 1979.

L.A. Real. The kinetics of functional response. *Am. Nat.*, 111: 289–300, 1977.

T. Reichenbach, M. Mobilia, and E. Frey. Mobility promotes and jeopardizes biodiversity in rockpaperscissors games. *Nature*, 448: 1046–1049, 2007.

E. Renshaw. *Modelling Biological Populations in Space and Time*. Cambridge University Press, Cambridge, 1991.

O. Reynolds. An experimental investigation of the circumstances which determine whether the motion of water shall be direct or sinuous and the law of resistance in parallel channels. *Phil. Trans. R. Soc.*, 174: 935–982, 1883.

O. Reynolds. On the dynamical theory of incompressible viscous fluids and the determination of the criterion. *Phil. Trans. R. Soc. Lond. A*, 186: 123–164, 1895.

L.F. Richardson. *Weather Prediction by Numerical Process*. Cambridge Univ. Press, 1922.

L.F. Richardson. Atmospheric diffusion shown on a distance-neighbour graph. *Proc. Roy. Soc. A*, 110: 709–737, 1926.

M.N. Rosenbluth, H.L. Berk, I. Doxas, and W. Horton. Effective diffusion in laminar convective flows. *Phys. Fluids*, 30: 2636–2647, 1987.

D. Rothstein, E. Henry, and J.P. Gollub. Persistent patterns in transient chaotic fluid mixing. *Nature*, 401: 770–772, 1999.

A.B. Rovinsky, H. Adiwidjaja, V.Z. Yakhnin, and M. Mezinger. Patchiness and enhancement of productivity in plankton ecosystems due to differential advection of predator and prey. *Oikos*, 78: 101–106, 1997.

D. Saintillan and M.J. Shelley. Orientational order and instabilities in suspensions of self-locomoting rods. *Phys. Rev. Lett.*, 99: 058102, 2007.

D. Saintillan and M.J. Shelley. Instabilities and pattern formation in active particle suspensions: Kinetic theory and continuum simulations. *Phys. Rev. Lett.*, 100: 178103, 2008.

M. Sanati and A. Saxena. Domain walls in [omega]-phase transformations. *Physica D*, 123: 368–379, 1998.

I. Scheuring, G. Károlyi, Á. Péntek, T. Tél, and Z. Toroczkai. A model for resolving the plankton paradox: coexistence in open flows. *Freshwater Biol.*, 45: 123–132, 2000.

I. Scheuring, G. Károlyi, Z. Toroczkai, T. Tél, and Á. Péntek. Competing populations in flows with chaotic mixing. *Theor. Pop. Biol.*, 63: 77–90, 2003.

S.K. Scott. *Chemical Chaos*. Oxford University Press, Oxford, 1991.

L.A. Segel and M. Slemrod. The quasi-steady state assumption: A case study in perturbation. *SIAM Review*, 31: 446–477, 1989.

M. Shibata and J. Bureš. Optimum topographical conditions for reverberating cortical spreading depression in rats. *J. Neurobiol.*, 5: 107–118, 1974.

B.I. Shraiman. Diffusive transport in a Rayleigh-Bénard convection cell. *Phys. Rev. A*, 36: 261–267, 1987.

B.I. Shraiman, A. Pumir, W. van Saarloos, P.C. Hohenberg, H. Chaté, and M. Holen. Spatiotemporal chaos in the one-dimensional complex Ginzburg-Landau equation. *Physica D*, 57: 241–248, 1992.

B.I. Shraiman and E.D. Siggia. Scalar turbulence. *Nature*, 405: 639–646, 2000.

H. Sigurgeirsson and A.M. Stuart. A model for preferential concentration. *Phys. Fluids*, 14: 4352, 2002.

B. Sinervo and M. Lively. The rock-paper-scissors game and the evolution of alternative male strategies. *Nature*, 380: 240–243, 1996.

J.G. Skellam. Random dispersal in theoretical populations. *Biometrika*, 38: 196, 1951.

A. Sokolov, I.S. Aranson, J.O. Kessler, and R.E. Goldstein. Concentration dependence of the collective dynamics of swimming bacteria. *Phys. Rev. Lett.*, 98: 158102, 2007.

I.M. Sokolov and A. Blumen. Mixing effects in the $A + B \to 0$ reaction-diffusion scheme. *Phys. Rev. Lett.*, 66: 1942–1945, 1991.

T.H. Solomon and J.P. Gollub. Passive transport in steady Rayleigh-Bénard convection. *Phys. Fluids*, 31: 1372–1379, 1988.

T.H. Solomon and I. Mezic. Uniform resonant chaotic mixing in fluid flows. *Nature*, 425: 376–380, 2003.

T.H. Solomon, E.R. Weeks, and H.L. Swinney. Observation of anomalous diffusion and Lévy flights in a two-dimensional rotating flow. *Phys. Rev. Lett.*, 71: 3975–3978, 1993.

T.H. Solomon, E.R. Weeks, and H.L. Swinney. Chaotic advection in a two-dimensional flow: Lévy flights and anomalous diffusion. *Physica D*, 76: 70–84, 1994.

J.C. Sommerer, H.-C. Ku, and H.E. Gilreath. Experimental evidence for chaotic scattering in a fluid wake. *Phys. Rev. Lett.*, 77: 5055–5058, 1996.

T.M. Squires. Microfluidics: Fluid physics at the nanoliter scale. *Rev. Mod. Phys.*, 77: 977–1026, 2005.

K.R. Sreenivasan and R.A. Antonia. The phenomenology of small-scale turbulence. *Ann. Rev. Fluid Mech.*, 29: 435–472, 1997.

M.A. Srokosz, G.D. Quartly, and J. J. H. Buck. A possible plankton wave in the Indian Ocean. *Geophys. Res. Lett.*, 31: L13301, 2004.

A.V. Straube, M. Abel, and A. Pikovsky. Temporal chaos versus spatial mixing in reaction-advection-diffusion systems. *Phys. Rev. Lett.*, 93: 174501, 2004.

A.D. Stroock, S.K.W. Dertinger, A. Ajdari, I. Mezic, H.A. Stone, and G.M. Whitesides. Chaotic mixer for microchannels. *Science*, 295: 647, 2002.

J. Sukhatme and R.T. Pierrehumbert. Decay of passive scalars under the action of single scale smooth velocity fields in bounded two-dimensional domains: From non-self-similar probability distribution functions to self-similar eigenmodes. *Phys. Rev. E*, 66: 056302, 2002.

E.S. Szalai, J. Kukura, P.E. Arratia, and F.J. Muzzio. Effects of hydrodynamics on reactive mixing in laminar flows. *AIChE J.*, 49: 168–179, 2003.

P. Tabeling. Two-dimensional turbulence: A physicist approach. *Phys. Rep.*, 362: 1–62, 2002.

G.I. Taylor. Diffusion by continuous movements. *Proc. Roy. Soc. London A*, 110: 709, 1921.

G.I. Taylor. Dispersion of soluble matter in solvent flowing through a tube. *Proc. Roy. Soc. London*, A219: 186–203, 1953.

T. Tél, A. de Moura, C. Grebogi, and G. Károlyi. Chemical and biological activity in open flows: A dynamical system approach. *Phys. Rep.*, 413: 91–196, 2005. Erratum: Phys. Rep. 415, 360 (2005).

J.-L. Thiffeault. Scalar decay in chaotic mixing. In J.B. Weiss and A. Provenzale, editors, *Transport and Mixing in Geophysical Flows*, volume 744 of *Lecture Notes in Physics*, pages 3–36. Springer Verlag, Berlin/Heildeger, 2008.

J.-L. Thiffeault and S. Childress. Chaotic mixing in a torus map. *Chaos*, 13: 502–507, 2003.

V.M. Tikhomirov. *Selected works of A.N. Kolmogorov*. Kluwer Academic Publishers, London, 1991.

Y. Togashi and K. Kaneko. Transitions induced by the discreteness of molecules in a small autocatalytic system. *Phys. Rev. Lett.*, 86 (11): 2459–2462, 2001.

C. Torney and Z. Neufeld. Transport and aggregation of self-propelled particles in fluid flows. *Phys. Rev. Lett.*, 99: 078101, 2007.

C. Torney and Z. Neufeld. Phototactic clustering of swimming microorganisms in a turbulent velocity field. *Phys. Rev. Lett.*, 101: 078105, 2008.

Z. Toroczkai, G. Károlyi, A. Péntek, T. Tél, and C. Grebogi. Advection of active particles in open chaotic flows. *Phys. Rev. Lett.*, 80: 500–503, 1998.

Z. Toroczkai and T. Tél, editors. *Focus Issue on Active Chaotic Flow*, volume 12, pages 372–530. 2002.

D.J. Tritton. *Physical Fluid Dynamics*. Oxford University Press, Oxford, 1988.

J.E. Truscott and J. Brindley. Ocean plankton populations as excitable media. *Bull. Math. Biol.*, 56: 981–998, 1994.

Y.-K. Tsang, T.M. Antonsen, and E. Ott. Exponential decay of chaotically advected passive scalars in the zero diffusivity limit. *Phys. Rev. E*, 71: 066301, 2005.

A.M. Turing. The chemical basis of morphogenesis. *Phil. Trans. Roy. Soc. Lond. B*, 237: 37–72, 1952.

I. Tuval, L. Cisneros, C. Dombrowski, C.W. Wolgemuth, J. Kessler, and R.E. Goldstein. Bacterial swimming and oxygen transport near contact lines. *Proc. Natl. Acad. Sci. USA*, 102: 2277–2282, 2005.

J. Tyson. A quantitative account of oscillations, bistability, and travelling waves in the Belousov-Zhabotinskii reaction. In *Field and Burger (1985)*, pages 92–144. 1985.

N.G. van Kampen. *Stochastic Processes in Physics and Chemistry*. North-Holland, Amsterdam, 1992.

W. van Saarloos. Front propagation into unstable states. *Phys. Rep.*, 386: 29–222, 2003.

J.C. Vassilicos and J.C.R. Hunt. Fractal dimensions and spectra of interfacs with applications to turbulence. *Proc. Roy. Soc. Lond. A*, 435: 505–534, 1991.

M. Vergassola, E. Villermaux, and B.I. Shraiman. 'Infotaxis' as a strategy for searching without gradients. *Nature*, 445: 406–409, 2007.

A. Vikhansky and S. M. Cox. Conditional moment closure for chemical reactions in laminar chaotic flows. *AIChE J.*, 53: 19–27, 2007.

A. Vikhansky and S.M. Cox. Reduced models of chemical reaction in chaotic flows. *Phys. Fluids*, 18: 037102, 2006.

V. Volterra. Variazioni e fluttuazioni del numero d'individui in specie animali conviventi (Variations and fluctuations of the number of individuals in animal species living together). *Memoria della R. Accademia Nazionale dei Lincei, Ser. VI*, 2: 31–113, 1926. Translated in (Chapman, 1931), pp. 409–448.

D. Walgraef. *Spatio-Temporal Pattern Formation: With Examples from Physics, Chemistry, and Materials Science*. Springer-Verlag, Berlin, 1996.

P. Waltman. *Competition Models in Population Biology*. Society for Industrial and Applied Mathematics, Philadelphia, 1983.

E.R. Weeks, J.S. Urbach, and H.L. Swinney. Anomalous diffusion in asymmetric random walks with a quasi-geostrophic flow example. *PhysicaD*, 97: 291–310, 1996.

S. Wiggins and J.M. Ottino. Foundations of chaotic mixing. *Phil. Trans. Roy. Soc. London. A*, 362: 937–970, 2004.

B.S. Williams, D. Marteau, and J.P. Gollub. Mixing of a passive scalar in magnetically forced two-dimensional turbulence. *Phys. Fluids*, 9: 2061, 1997.

A. Wonhas and J.C. Vassilicos. Diffusivity dependence of ozone depletion over the midnorthern latitudes. *Phys. Rev. E*, 65: 051111, 2002.

X.-L. Wu and A. Libchaber. Particle diffusion in a quasi-two-dimensional bacterial bath. *Phys. Rev. Lett.*, 86: 556, 2000.

A.M. Zhabotinsky. Periodic processes of the oxidation of malonic acid in solution (Study of the kinetics of Belousov's reaction). *Biofizika*, 9: 306–311, 1964.

C. Zhou, J. Kurths, Z. Neufeld, and I.Z. Kiss. Noise-sustained coherent oscillation of excitable media in a chaotic flow. *Phys. Rev. Lett.*, 91: 150601, 2003.

E. Ziemniak, C. Jung, and T. Tél. Tracer dynamics in open hydrodynamical flows as chaotic scattering. *Physica*, D 76: 123–146, 1994. erratum in Physica D 79, (1994),424.

Index